Sustainable Chemistry

WIT*PRESS*

WIT Press publishes leading books in Science and Technology.
Visit our website for the current list of titles.
www.witpress.com

WIT*eLibrary*

Home of the Transactions of the Wessex Institute.
Papers presented at Sustainable Chemistry are archived in the WIT elibrary in volume 154 of WIT Transactions on Ecology and the Environment (ISSN 1743-3541).
The WIT electronic-library provides the international scientific community with immediate and permanent access to individual papers presented at WIT conferences.
http://library.witpress.com

First International Conference on
Sustainable Chemistry

Sustainable Chemistry 2011

Conference Chairmen

G. Reniers
University of Antwerp, Belgium

C.A. Brebbia
Wessex Institute of Technology, UK

International Scientific Advisory Committee

Organised by
Wessex Institute of Technology, UK
University of Antwerp, Belgium

Sponsored by
WIT Transactions on Ecology and the Environment
&
FISCH, Antwerp, Belgium

Universiteit
Antwerpen

WIT Transactions

Transactions Editor

Editorial Board

Sustainable Chemistry

Editors

G. Reniers
University of Antwerp, Belgium

C.A. Brebbia
Wessex Institute of Technology, UK

WIT*PRESS* **Southampton, Boston**

Editor:

G. Reniers
University of Antwerp, Belgium

C.A. Brebbia
Wessex Institute of Technology, UK

Published by

WIT Press
Ashurst Lodge, Ashurst, Southampton, SO40 7AA, UK
Tel: 44 (0) 238 029 3223; Fax: 44 (0) 238 029 2853
E-Mail: witpress@witpress.com
http://www.witpress.com

For USA, Canada and Mexico

Computational Mechanics Inc
25 Bridge Street, Billerica, MA 01821, USA
Tel: 978 667 5841; Fax: 978 667 7582
E-Mail: infousa@witpress.com
http://www.witpress.com

British Library Cataloguing-in-Publication Data

A Catalogue record for this book is available
from the British Library

ISBN: 978-1-84564-558-8
ISSN: (print) 1746-448X
ISSN: (on-line) 1743-3541

The texts of the papers in this volume were set individually by the authors or under their supervision. Only minor corrections to the text may have been carried out by the publisher.

Printed in Great Britain by Quay Digital, Bristol.

Preface

Chemical products make an irreplaceable contribution in every aspect of our modern-day lives. Chemical processes and products play an essential role in industrial sectors as diverse as agriculture, automotive, clothing, communication, construction, food, health, leisure, mobility, plastics, space, transport, etc. We can easily observe that our advanced society depends on the wealth-creating aspects of industrial chemistry.

Societal expectations and the depletion of natural resources are pushing towards chemical processes becoming cleaner, more efficient, less consuming, safer and more secured. The ecological footprint of chemical products needs to be decreased.

Sustainable chemistry being concerned with the development of sustainable chemical products and processes and thereby integrating economic, environmental and social performance, can provide an answer to these major challenges.

The technological and managerial objectives of scientific and industrial research and activities are quite straightforward; decrease the use of energy and fossil materials, develop clean, safe and secured processes, produce useful, safe and sustainable products, develop a long-term scope for transportation activities, etc., always taking factors such as cost-effectiveness, eco-efficiency and inherent safety into consideration.

A clear interdisciplinary approach within technological areas, supported by cross-cutting managerial actions, is required for successful tackling these new chemistry paradigms.

This book includes papers presented at the 1st International Conference on Sustainable Chemistry (CHEM 2011). The meeting provided a forum for the presentation and discussion of the most recent developments in the theoretical and practical aspects of sustainable chemistry. The topics covered by the various presenters included:

- Eco-efficiency
- Smart Processing technology for sustainability
- Improvement in catalysis
- Multifunctional materials
- Bio-based materials
- Environmental heath issues

The Editors would like to express their appreciation of the valuable advice they received from the members of the International Scientific Advisory Committee and thank the contributors for the quality of their papers.

The Editors,
Antwerp, 2011

Contents

Section 5: Bio-based materials

Section 6: Environmental health issues

Section 1
Eco-efficiency

How to enhance sustainable chemistry in a non-technological way?

G. L. L. Reniers[1,2] & K. Sörensen[1]
[1]*Faculty of Applied Economic Sciences, Research Groups ARGoSS and ANT/OR, University of Antwerp, Belgium*
[2]*Centre for Economics and Corporate Sustainability (CEDON), Belgium*

Abstract

To achieve sustainable industrial chemical processes and products, companies, research centers and academia tend to focus mainly on technological solutions such as cleantech, green technology, process intensification, new catalysts, new membranes, ecofining, etc. However, non-technological approaches are essential as well to succeed in adequate sustainable chemistry. Cluster management, sustainable supply chain management, chemical leasing, integrated management systems and business models, societal expectations, etc. are all important non-technological aspects of sustainable chemistry. To date, most of the know-how and expertise on non-technological issues is developed on an individual company or academia basis and in a fragmented way. It is nonetheless crucial for the vision of sustainable chemistry to be realized that not only novel technology is conceptualized and developed by individual initiatives, but also that innovative management models, intra-organization models, and inter-organization models are elaborated, promoted and implemented by academia and by industry on a much broader scale.
Keywords: non-technological sustainable chemistry, chemical leasing, integrated management systems.

1 Introduction

According to the definition of Brundtland, sustainable development is the development that fulfils the needs of the present generation without compromising the ability of future generations to meet their own needs [1]. This means that long-term thinking and acting is needed. As the advantages of a long-

WIT Transactions on Ecology and the Environment, Vol 154, © 2011 WIT Press
www.witpress.com, ISSN 1743-3541 (on-line)
doi:10.2495/CHEM110011

term approach are not always directly recognized as profitable on the short-term, it is essential that a constant overview of the value chain is installed in a company and that the opportunities that can be found in such a long-term approach are recognized. Value chain opportunities such as chemical leasing, organized supply chains and transport, waste exchange, cradle-to-cradle actions, integrated SHESQ management systems or strategic collaboration may be elaborated within organizations to anchor sustainability as a vision.

Managers are discovering that the indicators that gauge sustainability can also be indicators of efficacy – that is, how well a company is *run*. From the management of corporate liabilities to new market ventures, a sustainable business strategy can improve all segments of corporate activity. It can be argued that management with focus on sustainability is a good proxy for gauging overall management capabilities at strategic, tactic and operational levels.

Traditionally, sustainability issues such as environmental efforts and social welfare expenditures have been viewed as costs that correlate negative with returns. However, studies suggest that there are several opportunities for competitive advantage and increased profits to be gained by strategic sustainability initiatives [2].

This reasoning reflects a shift from viewing business expenditures in a static world, to viewing them in a dynamic one based on innovation. Porter and Van der Linde [3] argue that in static model, firms and clusters of firms have already made their cost-minimizing choices and therefore any imperative to spend 'in the name of sustainability' inevitably raises costs. Such a static world falsely assumes that profit-seeking in se leads automatically to the pursuit of all profitable innovations. However, a dynamic world, which is actually the real world that companies operate in, is shaped by the stimulation and development of innovations. In other words, managing with focus on sustainability, that is, enhancing sustainability in a non-technological way, have helped spark (non-technological as well as technological) innovations that have eventually improved chemical process efficiencies, tapped new markets, streamlined productions and materials use, reduced pollution, and led to many other benefits.

Nonetheless, public and private researchers still need to bestow much greater effort in studying essential non-technological domains for improving sustainability in the chemical industry and achieving sustainable chemical products and processes. This paper deals with the different fields of non-technological research deserving more attention by academia and industry.

If supply chain management is aimed at installing beneficial partnerships and seamless linkages between multiple parties operating at different levels of the supply chain to avoid unnecessary logistics costs, it is referred to as vertical logistics collaboration. Apart from the well-established concept of vertical collaboration, horizontal collaboration can be distinguished. Horizontal collaboration is used to refer to concerted practices to "share private information, facilities or resources to reduce costs or improve service between companies (competing or unrelated) operating at the same level(s) in the market" [4].

It is obvious that a chemical enterprise has several possible ways of installing organizational improvements for its aim to optimize chemical process or product sustainability. Figure 1 displays the different perspectives.

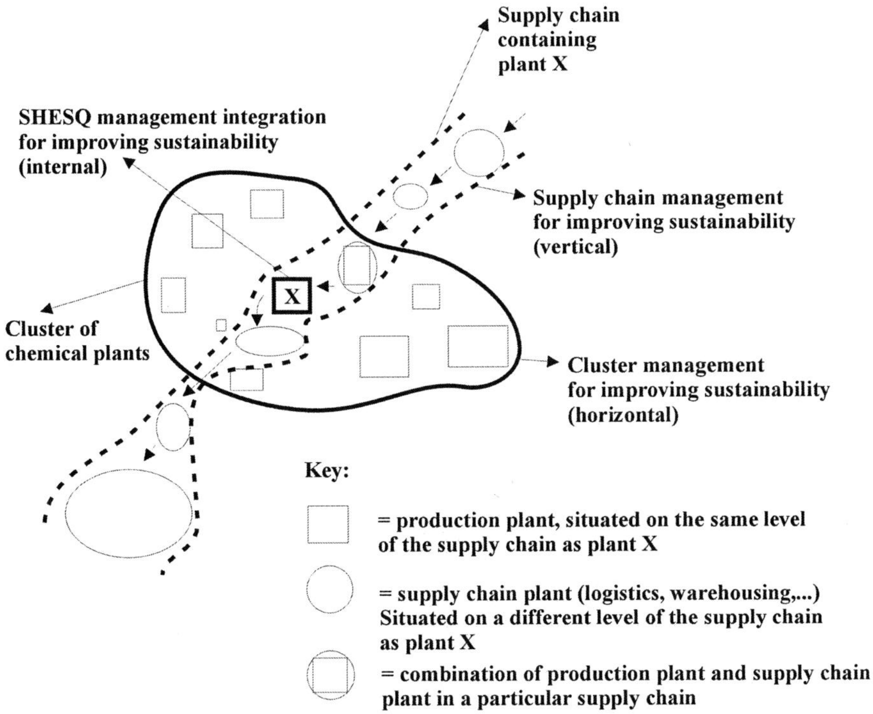

Figure 1: Different perspectives for achieving sustainable chemistry in a non-technological way.

As a first (horizontal) perspective, cross-plant management between corporations situated on the same level of the supply chain can be mentioned. This innovative perspective for enhancing sustainability in a non-technological way is called multi-plant management or cluster management.

As a second (vertical) perspective, supply chain management is envisioned. A supply chain policy trying to optimize sustainability within a chemical industrial area is aimed at realizing a more efficient chemical logistics chain.

The third (plant-internal) perspective is concerned with a company's safety, health, environment, security and quality management systems. By integrating these aforementioned management systems, chemical processes and products may be efficiently approached from a holistic people, planet and profit viewpoint.

The transition from a 'traditional' chemical plant towards a truly 'sustainable' corporation may only be achieved through the use of adequate business models directed at cooperation and multi-plant management. The three (non-technological) perspectives for elaborating these business models and leading towards the 'sustainable company' vision are discussed more in depth in the following Sections.

2 Cluster management

Although collaborative arrangements within many industries are well-known and often successful and appreciated, further optimization of these arrangements is many times possible. By augmenting collaborative agreements and relationships and by linking up with other firms on the same level of the market, a company may enjoy options otherwise unavailable to it, such as better access to markets, pooling or swapping of technologies and production volumes, access to specialized competencies, lower risk of R&D, enjoying larger economies of scale, benefiting from economies of scope, etc. [5, 6]. This way, collaborative arrangements often lead to more sustainable solutions and situations. For example, Seuring and Müller [7] discuss the cross-integration of integrated chain management and supply chain management from a conceptual viewpoint and describe five case-studies proving that collaborative arrangements are the only way to improve the competitiveness of the supply chain while reducing environmental burden.

The development of industrial eco-parks or industrial ecology is another example of how companies may improve their sustainable behavior through collaboration. Sterr and Ott [8] indicate that larger regional industrial areas may be more suitable for creating sustainable industrial ecosystems. At the same time, these authors also point out that the larger the industrial region, the more difficult to establish the necessary trust and coordination among the actors for setting up successful collaborative agreements.

To build stronger and more sustainability-oriented organizations, Lozano [9] suggests employing the Japanese concept of Kyosei, or "spirit of collaboration", complemented with a Multi-dimensional Sustainability Influence Change memework called MuSIC. This memework rightfully states that, in order to create an (ideal) efficient and effective paradigm shift towards sustainability within an organization, this change process should simultaneously happen in several dimensions, from individuals, groups and the organization through alignment and support by leadership and institutional frameworks, to the congruent change in informational, emotional and behavioural attitudes [9]. However, MuSIC does not include the inter-organizational dimension. Nonetheless, it is this very important inter-organizational dimension where a paradigm change process is needed for initiating collaboration between companies, which leads to truly sustainable industrial clusters. In our opinion, MuSIC should therefore be expanded with a cross-organizational dimension, providing a guide to understanding not only the intra-organizational interactions, but also the inter-organizational interactions, and helping support an organisation's as well as an industrial park's sustainability transformation.

A framework for allowing and enabling companies to take joint strategic decisions (e.g. investments) should be shaped, based on a multi-plant or cluster vision. Multi-plant opportunities with respect to innovativeness and sustainability should be identified and mapped.

An innovative cluster-leveled policy and the resulting innovative policies on individual companies' levels should lead to strategic decisions being implemented on tactic and operational levels, aiming to make chemical products and processes more sustainable.

Cluster management may also lead to the transition of managing safety, health, environment, security and quality within a group of chemical plants. By taking people (safety and security), planet (environment) and profit (quality) simultaneously into account in a cross-plant SHESQ management system/concept, improvement steps for individual plants may be achieved. Ethical aspects may also be introduced into this cross-plant management system and, by extension, to the individual plants. This way, a chemical industrial area may increase its efficiency, as well as greatly improve its internal and external multi-corporate image. Such an approach may also lead to an important competitive advantage for the individual companies participating into the cluster initiative.

3 Supply chain management

Another crucial factor in the development of sustainable chemistry is managing logistic processes within and between (chemical) companies in an optimal way. Obviously, technical innovations, such as process and product improvement and the replacement of fossil fuels with renewable raw materials, can deliver an important contribution to sustainability. However, the benefits of these innovations can be easily undone by an inefficient organization of the supply chain, both within a company and between different companies. To put it bluntly, a production process improvement that reduces CO_2 emissions by 20% can easily be undone if twice as many kilometers must be completed to deliver this product to the customers. Notwithstanding the evidence of logistic optimization, examples of major inefficiencies in the organization of logistics processes between companies (for example, large quantities of identical products are transported in both directions between chemical clusters) or within firms (e.g., large stocks resulting from a lack of demand forecasting) are all too common.

Logistics and supply chain management are not equivalent to physical distribution, but also include the sound management of the information necessary for an effective control of the supply chain. Where performance measures of supply chains in the past have been predominantly cost-based, a broad consensus has formed in the post-Kyoto period that logistics operations should proceed in a more sustainable way [10–12].Consequently, the optimization of the various components of the supply chain increasingly take into account criteria related to sustainability and the environment [13]. Simultaneously, various social criteria, such as working time restrictions, are also taken into account.

Sustainable optimization of the supply chain adds an additional level of complexity. Software for vehicle routing, scheduling and planning of the supply chain is therefore still based solely on cost. One reason is the lack of an overall methodology for making decisions involving multiple objectives simultaneously. Cost and sustainability are however both important, but conflicting criteria are present and a trade-off must be made between them.

Logistics in the chemical industry also differs significantly from logistics in other sectors and therefore requires a specific approach. This has an impact on many aspects of the supply chain such as inventory management, production management and distribution management. In the chemical industry transported products are often hazardous, which imposes important restrictions (permits, regulations) on the organization of the supply chain and possibly affects its flexibility. Another difference is that chemicals, unlike most other types of freight, can be transported by pipeline. This transport mode has some very specific properties, such as a lack of flexibility on the one hand, but a very small variable cost on the other hand. The fact that often large quantities of homogeneous products have to be carried makes the chemical industry an important subject for a clearing system. In such a system a competing supplier can perform the physical production and delivery of a generic product to a customer clients if this is advantageous (for example, because the competitor is located much closer to the customer). The competitor will then invoice the supplier for the services provided, or perform similar services and incur similar costs. In such a system, a supplier of ammonia in Houston can leave it to an Antwerp-based supplier to deliver to its customers in Antwerp and vice versa. In this way, unnecessary transportation and logistics operations are avoided. Evidently the success of such a system requires a high degree of openness between the various chemical companies.

4 SHESQ management integration

Operational risks may be classified into four distinct management domains: safety and health, security, environment, and quality. Management systems and norms are available to organizations in each of the four domains (e.g. ISO 9001:2000 (quality), OSHAS 18001:2007 (safety and health) en EMAS2, ISO 14000:2004 (environment)). The traditional approach of employing a different management system in every domain for dealing with operational risks in a company has resulted in parallel, and thus entirely separate, management systems. Every domain has its own history, specificities, and insights, such as company know-how (quality), expectations from the Government and other stakeholders (safety and environment), and societal evolutions (security). As such, it can be understood and explained that separate obligations, models and instruments have emerged in the different domains: domain-specific legislation, management models, risk analysis techniques, education and training sessions, distinct functions within companies (such as e.g. the prevention advisor, the environmental coordinator, the security liaison officer), etc. For every management domain, different performance indicators are used as well.

Nevertheless, public as well as private bodies recognize the need for an integrative approach to deal with the four domains, besides a specialist approach, and framed within the sustainability concept of organizations. Karapetrovic and Jonker [14] indicate that integrating standardized management systems leads to synergetic effects and significant savings in business operations. According to Smith [15], an integrated management system has the additional advantage that improvements are implemented simultaneously within all four domains. This way, an integrated management system works in a pro-active manner. Embedding sustainability concepts into the system results in a continuous improvement in all business operations (e.g. also in case of technological innovations or in case of supply chain optimizations).

The evolutionary process of a mono-disciplinary risk approach towards an integrative risk approach within organizations can be illustrated via the development of Corporate Social Responsibility, which is a non-technological business concept.

5 Conclusions

The objective of writing an article from a managerial viewpoint consists in leveraging the search for truly sustainable chemical products and processes, and to disseminate the non-technological ideas and concepts to captains of industry and to leaders of the public sector, as well as to company management (within all organizational levels and from all different departments and disciplines).

It is obvious that realizing sustainable chemistry strongly depends on the levels at which innovative technological breakthroughs are possible and develop. This article indicates that inter-organizational cluster policies, managing the chemical supply chain more efficiently and integrating SHESQ management systems within companies, should also be considered as essential non-technological ways to stimulate innovation and sustainable technology.

Adequate cross-company policies for example lead to more effective safety and security within industrial parks, as well as increased energy-efficiencies and less waste streams. Optimizing the chemical supply chain leads amongst others to economic and ecological more efficient product streams. Integrating management systems results e.g. in continuous improvement of the sustainable chemistry concept within corporations, leading to competitive advantages.

References

[1] United Nations (1987). Report of the World Commission on Environment and Development. General Assembly Resolution 42/187, 11 December 1987.

[2] Reinhardt, F.L. (1999). Bringing the environment down to Earth, Harvard Business Review, p. 149-157.

[3] Porter, M.E., van der Linde, C. (1995). Toward a new conception of the environment – Competitiveness Relationship, Journal of Economic Perspectives, 9(4), p. 97-118.

[4] European Union (2001). Guidelines on the applicability of Article 81 of the EC Treaty to horizontal collaboration agreements, European Commission Notice 2001/C 3/02.

[5] De Man, A., van der Zee, H., Geurts, D., Kuijt, M., Vincent, N. (2000). Competing for partners, Pearson Education, Zeist, The Netherlands, 211p.

[6] Contractor, F.J., Lorange, P. (2002). Cooperative Strategies in international business. Joint ventures and technology partnerships between firms. Pergamon, Amsterdam, The Netherlands.

[7] Seuring, S., Müller, M. (2008). From a literature review to a conceptual framework for sustainable supply chain management, *Journal of Cleaner Production*, 16, pp. 1699-1710.

[8] Sterr, T., Ott, T. (2004). The industrial region as a promising unit for eco-industrial development – reflections, practical experience and establishment of innovative instruments to support industrial ecology, *Journal of Cleaner Production*, 12, pp. 947-965.

[9] Lozano, R. (2008). Developing collaborative and sustainable organizations, *Journal of Cleaner Production*, 16, pp. 499-509.

[10] European Commission (2007). Freight transport logistics action plan. Communication of the Commission of the European Communities.

[11] Lambert, D.M., Cooper, M.C. (2000). Issues in supply chain management. Industrial Marketing Management, 29, p. 65–83.

[12] Lamming, R., Hampson J. (1996). The environment as a supply chain management issue. British Journal of Management, 7, S45–S62, 336.

[13] Christopher, M. (1999). Logistics and supply chain management: Strategies for reducing cost and improving service (2nd Ed.), International Journal of Logistics Research and Applications, 2, 103.

[14] Karapetrovic, S., Jonker, J. (2003). Integration of standardized management systems: searching for a recipe and ingredients. Total Quality Management, 14(4), p. 451-459.

[15] Smith, D. (2002). IMS: The Framework, London: British Standards Institute.

Comparison of two pharmaceutical production processes using different eco-efficiency measuring methods

G. Van der Vorst[1], W. Aelterman[2], P. Van Broeck[2], S. Walraedt[3], K. Schaerlaekens[3], P. Stouthuyzen[4], H. Van Langenhove[1] & J. Dewulf[1]
[1]*Ghent University, Belgium*
[2]*Janssen Pharmaceutica, Belgium*
[3]*Essenscia, Belgium*
[4]*VITO, Belgium*

Abstract

The rising importance of eco-efficiency in the chemical industry resulted in the development of a huge amount of measuring methods for the assessment of environmental and/or economical sustainability. In this presentation, a limited set of eco-efficiency measuring methods are illustrated for a case in the pharmaceutical industry. Different eco-efficiency measuring methods as, for example, exergy analysis, carbon footprint or ETH's Finechem tool, are used for the evaluation and comparison of a pharmaceutical batch production step and a continuous production step using a micro reactor. Data for both processes are delivered by Janssen Pharmaceutica (Belgium). First, this case allows one to make a comparative evaluation of the eco-efficiency of the pharmaceutical production options. Second, a thorough evaluation of the capabilities and advantages of the different eco-efficiency measuring methods can be made.

The evaluation of two pharmaceutical production alternatives based on different eco-efficiency measuring methods is a case study in the Eco²chem project. In this Eco²chem project a structured evaluation of different eco-efficiency measuring methods for the chemical industry is made. The result of this Eco²chem project will be a web based eco-efficiency decision matrix allowing chemical companies to choose those eco-efficiency measuring methods which best fit the companies' needs.
Keywords: eco²chem, eco-efficiency measuring methods, pharmaceutical production.

WIT Transactions on Ecology and the Environment, Vol 154, © 2011 WIT Press
www.witpress.com, ISSN 1743-3541 (on-line)
doi:10.2495/CHEM110021

1 Introduction

Currently, there is no need to convince the chemical industry of the necessity to shift towards more eco-efficient production processes and production technologies. Research in this area is twofold. First, research efforts are required concerning development of innovative chemical reactions and technologies. Second, there is a need for an adequate assessment of the eco-efficiency Van der Vorst et al. [4]. This assessment is necessary for better decision making, but it also allows better communication, be it for stimulating involvement of personnel or for external use. Assessing the eco-efficiency of processes and technologies can be done by different eco-efficiency measuring methods (EEMM's). A wide range of EEMM's are made available to the chemical industry by scientific institutes, academics as well as industry itself. It is the aim of the Eco²chem Project, funded by European Regional Development Fund (ERDF) and the Flemish government, to develop a web based decision tool to help the chemical industry to select the EEMM which best fits the companies needs. In order to build this tool, a structured inventory of the existing EEMMs is made and a number of EEMM's is evaluated on real cases provided by the chemical industry. A limited set of eco-efficiency measuring methods are illustrated here for a case in the pharmaceutical industry. Eco-efficiency measuring methods as for example exergy analysis, carbon footprint, ETH's Finechem tool, life cycle assessment, E-factor etc. [1–3, 5] are used for the evaluation and comparison of a pharmaceutical batch production step and a continuous production step using a micro reactor. Data for both processes are delivered by Janssen Pharmaceutica (Belgium) [6].

2 Materials and methods

2.1 Eco-efficiency measuring methods (EEMMs)

For the evaluation of the eco-efficiency of both pharmaceutical production processes (batch vs. continuous), different EEMMs will be used. It is the purpose to calculate the eco-efficiency by using a wide range of different EEMMs and evaluate these results. Using this set of EEMMs will allow better understanding of the principles and possibilities of the EEMMs under consideration. This will contribute to the EEMM inventory to be made during the Eco²chem project and finally resulting in the Eco²chem EEMM decision tool. The specific EEMMs used for this comparison (batch vs. continuous pharmaceutical production step) will be the E-factor, ETH's Finechem tool, exergy analysis at the process and the plant level, Cumulative Exergy Extracted from the Natural Environment (CEENE) method, carbon footprint (IPCC 2007 – GWP), Eco-indicator '99, ecological footprint and the cumulative energy demand (CED) [2–4, 7–10]. More information on EEMMs can be found in the references.

2.2 Pharmaceutical production processes: batch versus continuous production

The case supplied by Janssen Pharmaceutica is the comparison of two alternatives for the sixth production step in the galantamine (anti-Alzheimer medication) production route. This sixth production step originally is a batch based production step, but can be replaced by a continuous production step using a micro reactor. In Figs. 1 and 2, an overview is given of the eight steps required

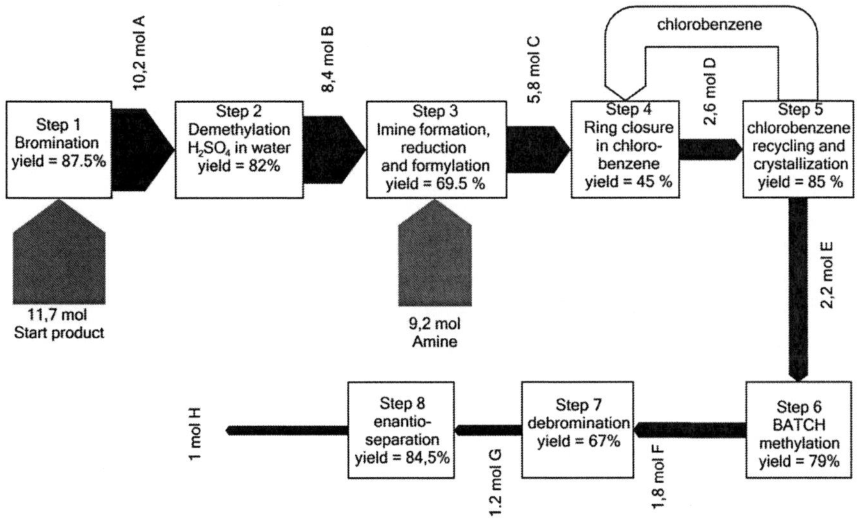

Figure 1: Synthesis route for the production of 1 mol intermediate H using the batch process in step 6.

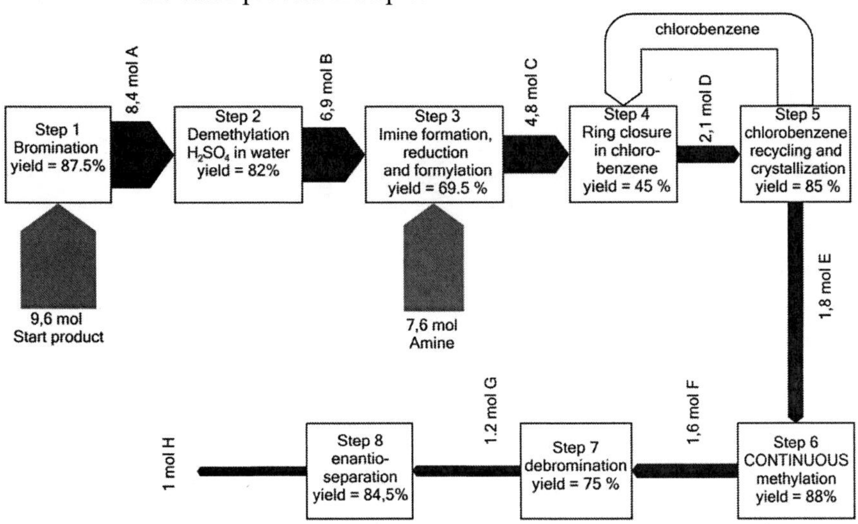

Figure 2: Synthesis route for the production of 1 mol intermediate H using the continuous process (micro reactor technology) in step 6.

for the production of 1 mol of the galantamine intermediate H. The evaluation by the different EEMMs is not limited to production step 6 but also shows the impact of taking into account the other steps of a pharmaceutical synthesis route. In Figs. 1 and 2, the improvement of the yield and its impact on the other production steps is illustrated.

The total data inventory required for the calculation of the used EEMMs, including all the mass and energy balances of all eight production steps is not given due to confidentiality issues and the overload of information.

3 Results

3.1 E-factor

The E-factor can be defined as the mass (kg) waste produced per kg product. In Fig. 3 the E-factor of the 8 production steps individually is visualized. This means that the waste produced in earlier production steps is not taken into account. Changing process step 6 from a batch into a continuous process results in a drop of the E-factor from 29kg waste/kg F to 19.5 kg waste/kg F. This corresponds to a reduction of almost 50%. In addition, small reductions due to the improved yields (see Figs. 1 and 2) can be seen in steps 7 and 8. However, Fig. 3 also illustrates that step 4 as is the production step with the highest E-factor. This E-factor is mostly covered by the high amount of wastewater produced as well by the low efficiency of the process.

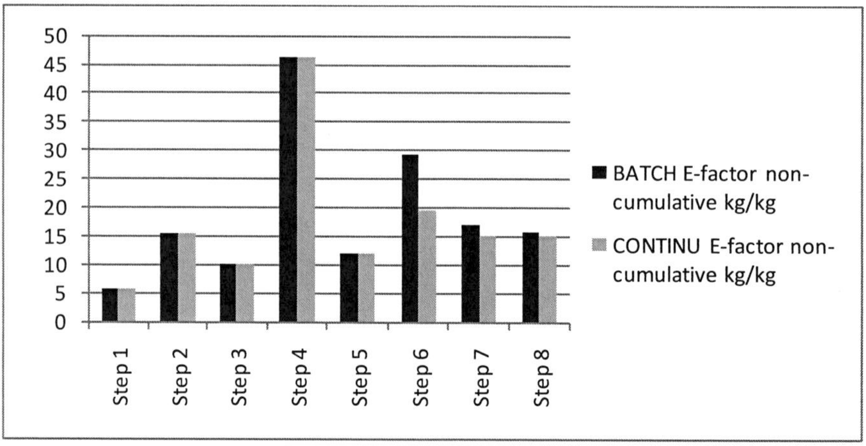

Figure 3: Non-cumulative E-factor of eight consecutive production steps.

In Fig. 4, the E-factors are cumulated. For each production step, the waste produced in the previous production step is taken into account. It can be seen that the reduction in the E-factor at step 6 by changing from batch to continuous is relatively small compared to the non-cumulative results. However, the difference increases again when taking into account steps 7 and 8. From these cumulative

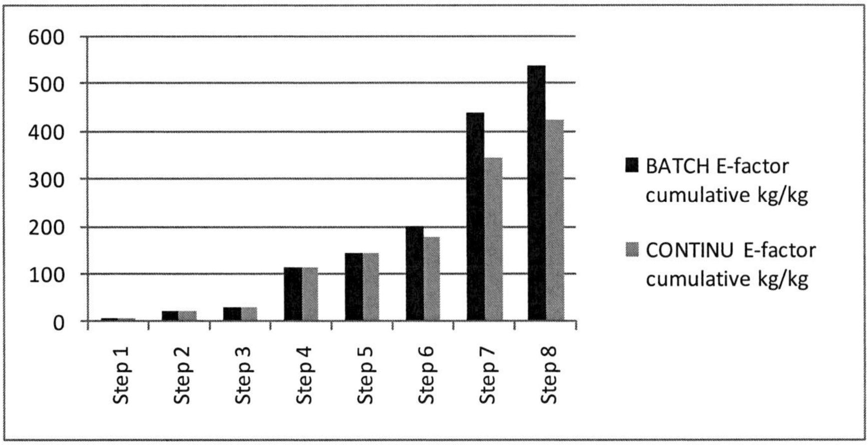

Figure 4: Cumulative E-factor of eight consecutive production steps.

results, it can be stated that in total 26% less waste is produced by using the continuous alternative in step 6.

3.2 ETH Finechem tool

The Finechem tool from ETH is an estimation tool for the prediction of the cumulative energy demand (CED), the global warming potential (GWP) and the eco-indicator '99 (EI99) based on the group contributions of the chemicals under consideration. This tool cannot be used for the estimation of the life cycle impact assessment (LCIA) of enantiomers and components containing bromine atoms. This means it is not a useful EEMM for the evaluation of this case. This is also clear from the results presented in Fig. 5 where the tool is used for illustration. From step 5 (molecule E) on, the environmental impacts do not increase anymore, which is impossible regarding the efficiencies in Figs. 1 and Fig. 2. The ETH Finechem tool however remains a very good estimation tool if no other data is available and as long as the guidelines are followed correctly, which is clearly not the case for this illustration.

3.3 Exergy analysis (process and plant level)

Next to the relatively quick EEMMs (E-factor and ETH finechem tool), more detailed but also more time consuming EEMMs can be used for the evaluation of chemical production processes. One example is the exergy analysis of the eight process steps at the process level and at the plant level. The focus here will only be on the results of plant level exergy analysis. Non-cumulative results at the plant level are presented in Fig. 6 and cumulative results are presented in Fig. 7. Those figures are similar to the ones presented for the E-factor (Figs. 3 and 4) because the main contributor of all the environmental impacts in these processes is the use of fossil chemicals (visible in Figs. 6 and 7). However, when Figs. 3 and 6 are put next to each other, the importance of step 4 in Fig. 3 has

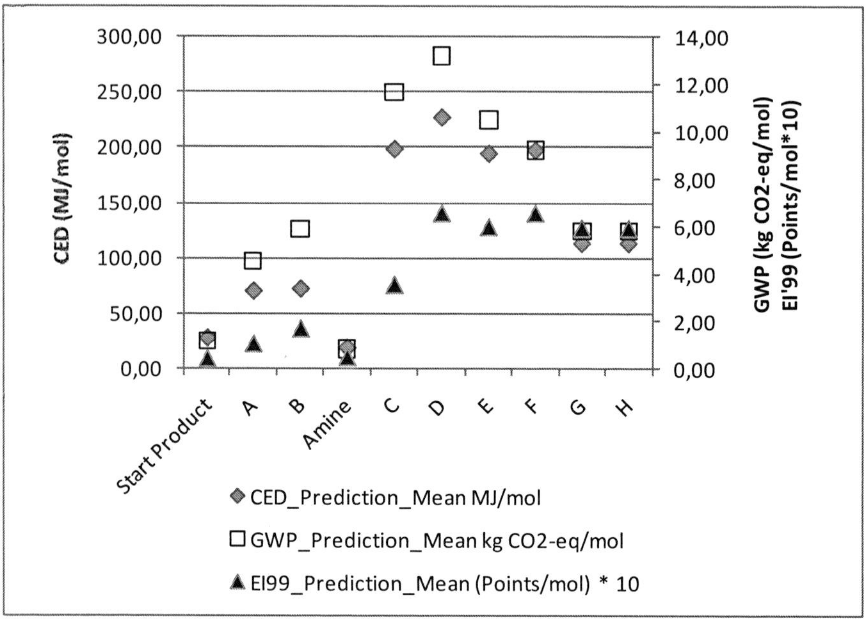

Figure 5: ETH prediction of ten intermediate molecules in the synthesis route of galantamine.

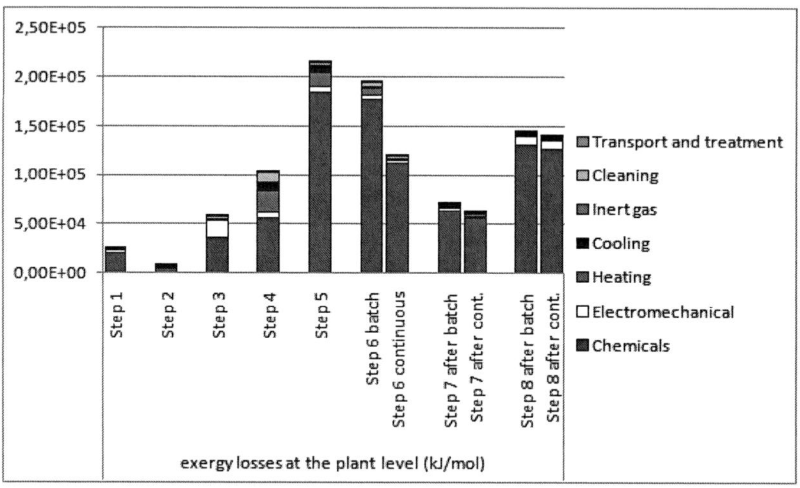

Figure 6: Non cumulative exergy losses at the plant level for eight consecutive production steps and divided over seven impact categories.

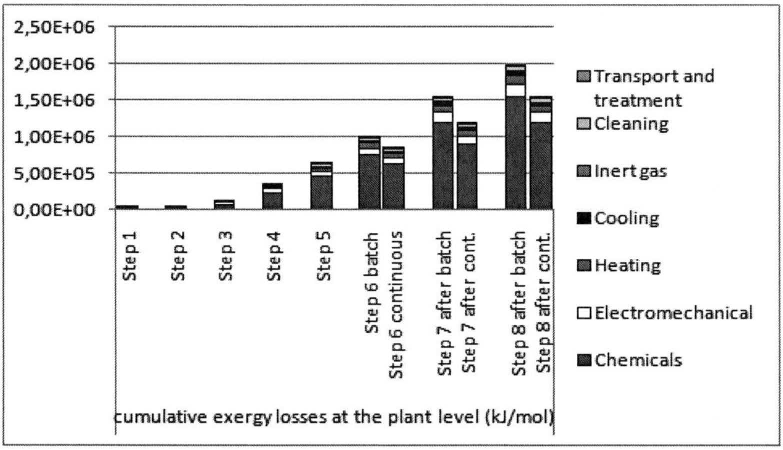

Figure 7: Cumulative exergy losses at the plant level for eight consecutive production steps and divided over seven impact categories.

disappeared in Fig. 6. The reason is that in the E-factor EEMM, one kg water has the same impact as one kg organic solvent. This is not the case using exergy analysis. The waste stream of step 4 is mainly water based and thus scores worse for the E-factor than for an exergy analysis. Coming back to the comparison of batch and continuous however, in Figs. 6 and 7, again the improvement of changing the process is clear and lies in the same order of magnitude as for the E-factor EEMM results.

3.4 Data intensive life cycle based evaluations (CEENE, CED, EI'99, IPCC 2007, EF)

The last evaluated EEMMs are grouped as life cycle based EEMMs and this includes EEMMs taking into account the full cradle-to-gate of the pharmaceutical production steps. Taking into account the full cradle-to-gate means more intensive data inventory is required. Similar as the results of the exergy analysis at the plant level, results can be presented (Fig. 8) by using the CEENE method at the cradle-to-gate level. In the exergy analysis at the plant level (Fig. 7), the resource consumption (exergy losses) were attributed to the sinks were they are used (lost). In the CEENE method, however, the resource consumption can also be attributed to the source the resources are coming from. As stated before, the highest impacts in Fig. 7 are linked to the use of fossil chemicals. This is confirmed in Fig. 8. The four other life cycle based EEMMs evaluated here show similar profiles with similar ratios between the process steps as presented in Fig. 8. In Table 1, the results of all the life cycle based EEMMs are presented and the improvements made by changing from batch to continuous production is given for the cumulative results of 1 mol F (stopping the evaluation after step 6) as well as for the cumulative results of 1 mol H (stopping the evaluations after step 8). First, it is clear that the improvements expressed in

Table 1: Impact reductions at step 6 and at step 8 for the five life cycle impact assessment methods.

	BATCH 1 mol F	CONTINU 1 mol F	Impact reduction	BATCH 1 mol H	CONTINU 1 mol H	Impact reduction
CEENE (kJ)	2,97E+06	2,54E+06	16,75%	5,64E+06	4,39E+06	28,54%
Carbon footprint IPCC GWP 2007 100a (kg CO2-eq)	9,99E+01	8,39E+01	19,10%	1,88E+02	1,43E+02	31,24%
Cumulative energy demand (total) (MJ-eq)	2,48E+03	2,12E+03	16,96%	4,71E+03	3,65E+03	28,89%
Ecoindicator EI'99 (H/A) total (points)	1,01E+01	8,61E+00	17,03%	1,88E+01	1,46E+01	29,33%
Ecological footprint (total) (m^2a)	2,66E+02	2,25E+02	18,40%	5,03E+02	3,86E+02	30,41%

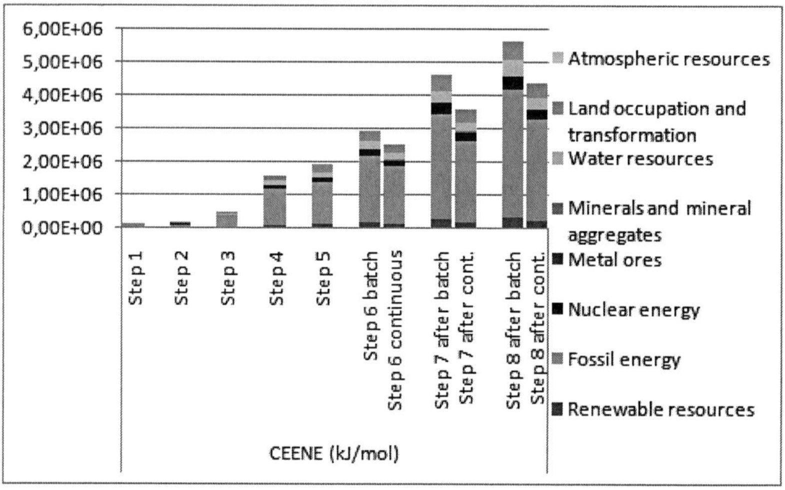

Figure 8: CEENE of the eight production steps (cumulative results).

percentages are similar for all 5 life cycle impact assessment methods. Reason is the use of organic solvents in all production steps. Second, it is clear that the improvements quantified in percentages can change significantly (from 18% up to 30%) if more consecutive production steps (step 6 up to step 8) are taken into account. This is related to the cumulative effect of taking into account the yields of the consecutive production steps.

4 Conclusions and outlook

Regarding the pharmaceutical production process evaluation, it can be concluded that for all the EEMMs used, the continuous alternative is better from an eco-efficiency point of view than the batch production process. Improvements ranging from 16 up to 50% are quantified depending on the used EEMM and the used boundaries (cumulative, non cumulative, process level, plant level, cradle-to-gate level).

Regarding the evaluation of the different used EEMMs, many different results can be obtained depending on the used eco-efficiency methodology. Although in this case all results were in favour of the continuous alternative, it can happen that for other cases the results will not be that obvious. Therefore it remains important to use the best EEMM for the purpose the user wants to have results for. Quick scans as the E-factor require less input which results in higher uncertainty on the results. The more detailed EEMMs are on the other hand more time-consuming. A good EEMM selection has to take into account these issues. It is also important to thoroughly study the limitations of different EEMM. If EEMMs (e.g. ETH finechem tool) limitations are not considered, big errors can be made without knowing.

Regarding the further outlook of the Eco²chem project, more cases are to be calculated for a better evaluation of a wider range of EEMMs. These results will finally contribute to a good evaluation and better understanding of the capabilities and limitations of the inventoried EEMMs. This knowledge will be applied in the development of the decision tool.

References

[1] Dewulf, J., Van Langenhove, H., Muys, B., Bruers, S., Bakshi, B.R., Grubb, G.F., Paulus, D.M. and Sciubba, E., Exergy: Its potential and limitations in environmental science and technology. *Environmental Science & Technology*, **42**, pp. 2221-2232, 2008.

[2] BSI , PAS 2050 (2008) Specification for the assessment of the life cycle greenhouse gas emissions of goods and services, www.bsigroup.com /Standards-and-Publications/How-we-can-help-you/Professional-Standards-Service/PAS-2050

[3] ETH Finechem tool, www.sust-chem.ethz.ch/tools/finechem/

[4] Van der Vorst, G., Van Langenhove, H., De Paep, F., Aelterman, W., Dingenen, J., Dewulf, J., Exergetic Life Cycle analysis for the selection of chromatographic separation processes in the pharmaceutical industry: preparative HPLC versus preparative SFC. *Green Chemistry*, **11**, pp. 1007-1012, 2009.

[5] European Commission – Joint Research Center, Life Cycle Thinking and Assessment, http://lct.jrc.ec.europa.eu/

[6] Janssen Pharmaceutica, www.janssenpharmaceutica.be

[7] Constable, D., Curzons, A., Cunningham, V., Metrics to 'green' chemistry – Which are the best? *Green Chemistry*, **4**, pp. 521-527, 2002.

[8] Dewulf, J., Bosch, M.E., De Meester, B., Van der Vorst, G., Van Langenhove, H., Hellweg, S., Huijbregts, M., Cumulative exergy extraction from the natural environment (CEENE): a comprehensive life cycle impact assessment method for resource accounting. *Environmental Science & Technology*, **41(24)**, pp. 8477-8483, 2007.

[9] Huijbregts, M., Hellweg, S., Frischknecht, R., Hungerbuehler, K., Hendriks, A., Ecological footprint accounting in the life cycle assessment of products. *Ecological Economics*, **64 (4)**, pp. 798-807, 2008.

[10] Huijbregts, M., Rombouts, L., Hellweg, S., Frischknecht, R., Hendriks, A., Van de Meent, D., Ragas, A., Reijnders, L., Struijs, J., Is cumulative fossil energy demand a useful indicator for the environmental performance of products, *Environmental Science & Technology*, **40(3)**, 641-648, 2006.

Solvents for sustainable chemical processes

P. Pollet[1,2], C. A. Eckert[1,2,3] & C. L. Liotta[1,2,3]
[1]*School of Chemistry & Biochemistry,*
Georgia Institute of Technology, USA
[2]*Specialty Separations Center, Georgia Institute of Technology, USA*
[3]*School of Chemical & Biomolecular Engineering,*
Georgia Institute of Technology, USA

Abstract

Worldwide, socioeconomic strategies encourage – and may soon enforce – industrial sustainability. Therefore, innovative research and approaches must provide short and long term solutions to reach future legislative targets. For chemical processes, the development of solvents that facilitate reaction and subsequent product separation is paramount to secure economic competitivity while minimizing impact on the environment and energy consumption. Tunable and switchable solvents were developed to address synergistically reaction and separation. Tunable solvents change properties *continuously* upon application of an external stimulus. For example, organic aqueous solvents allow for the reaction to be homogeneously catalyzed followed by a simple and efficient heterogeneous separation of the product (commonly a heterogeneous catalysis attribute). Chemical processes mediated by water at or below the near critical range can also alleviate the shortcomings of current synthetic strategies (waste production and management, remediation costs). In contrast, switchable solvents change physical properties *abruptly* upon application of an external stimulus. Piperylene sulfone is one example and can provide a recyclable alternative for dipolar, aprotic solvents such as dimethylsulfoxide (DMSO). Solvents can accomplish more than heat and mass transfer; they can actively contribute to facilitate reaction and product separation while minimizing waste generation and energy consumption.
Keywords: solvents, sustainability, reaction, separation, smart solvents, switchable, tunable, near critical water, piperylene sulfone, OATS.

1 Introduction

The solvents discussed here provide simple and efficient vehicles for conducting reactions and separations. This is accomplished by exposing the solvent system to an external physical or chemical stimulus which results in a dramatic change in its physical and chemical properties. Tunable solvents are defined as solvents that change properties continuously upon application of an external stimulus. Supercritrical fluids (SFC), (Dillow *et al.* [1], Brown *et al.* [2], Thompson *et al.* [3], Brown *et al.* [4], Nolen *et al.* [5], Eckert *et al.* [6], Koch *et al.* [7], Furstner *et al.* [8], Solinas *et al.* [9], Maayan *et al.* [10]) nearcritical liquids (Eckert *et al.* [6], Chandler *et al.* [11], Chandler *et al.* [12], Lesutis *et al.* [13], Patrick *et al.* [14], Nolen *et al.* [15]) and gas-expanded liquids (GXLs) (Jessop and Subramaniam [16], Ablan *et al.* [17], Jessop *et al.* [18]) are examples of tunable solvents. It has been demonstrated that these solvents systems are useful for coupling reactions and separations. (Dillow *et al.* [1], Eckert *et al.* [6], Daintree *et al.* [19], Eckert *et al.* [20], Eckert and Chamber [21], Eckert *et al.* [22, 23], Jessop [24], Ramsey *et al.* [25], Rayner [26], Rezaei *et al.* [27], Tang *et al.* [28]) In contrast, switchable solvents are solvents that change physical properties abruptly. In other words, they can be switched "on" and "off". This unique property is a consequence of a reversible reaction (i.e. addition/elimination reactions) in response to an external stimulus such as adjusting temperature and/or the addition or removal of a gas. Because of the reversibility of the reaction, the changed solvent can be brought back to its original state.

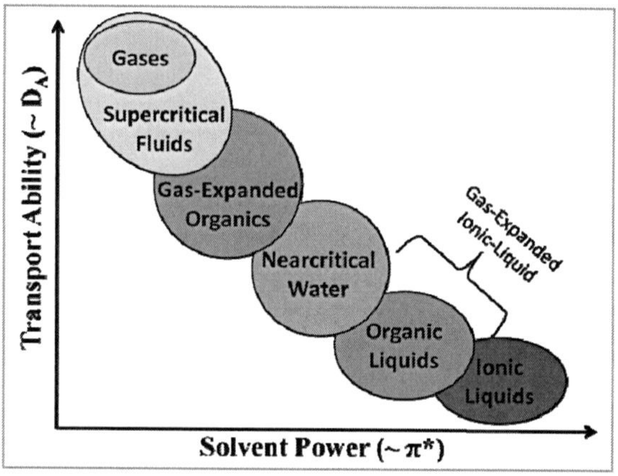

Figure 1: Solvent power and transport ability of various solvents.

Figure 1 relates qualitatively the solvent power (in terms of the Kamlet-Taft polarizability/dipolarity, π^*) to transport ability (in terms of diffusion coefficient, D_A) for the relevant solvents. Most liquids are strong solvents but they have low diffusion coefficients (about four to six orders of magnitude smaller than those for gases, Hines and Maddox [29], Bird *et al.* [30]) which may lead to mass transfer limitations. In contrast, gases have far better transport properties but are much weaker solvents. Supercritical fluids such as CO_2 are much stronger solvents than gaseous CO_2 and have much high diffusion coefficients than liquids. Gas-expanded liquids (GXLs), mixtures of organics and dissolved gases, are stronger solvents than supercritical fluids and have better diffusion coefficients than liquids. For any given process, the choice of solvent system depends on the physical and chemical properties of the reactants and products, the processing requirements, and environmental considerations.

It is important to develop reaction processes in a holistic manner. Just having a high yield reaction is not enough. The ease of separation of the desired product from the reaction system is also of paramount importance. As a consequence, our focus is placed on solvent systems that combine the benefits of homogeneous reactions and heterogeneous separations. Homogeneous reactions are usually superior to heterogeneous reactions in terms of reaction rates, process control, and selectivity. However, the difficulty of separating and recycling the catalyst limits economic applications of homogeneous catalysis. Heterogeneous catalysis is widely used industrially since it has a built-in separation of the product from the catalyst. The alternative solvent systems combine homogeneous reactions with heterogeneous separations. Herein, we discuss applications benefiting from the use of tunable and switchable solvents.

2 Tunable solvents

2.1 Organic aqueous tunable solvent systems (OATS)

Organic Aqueous Tunable Solvent (OATS) are homogeneous mixtures of aprotic organics (acetonitrile or tetrahydrofuran) and polar protic solvents (water or polyethylene glycol) that undergo a phase split to form biphasic liquid-liquid mixtures upon the addition of an antisolvent gas. The phase splitting – going from monophasic to biphasic – results from the difference in the antisolvent gas solubility between the aprotic organic solvent and the polar protic solvent. CO_2 is completely miscible with most organics but has only slightly solubility in aqueous media, and, as a consequence, is an effective antisolvent in promoting phase splitting. The resulting biphasic system consists of a GXL and a polar liquid phase. The physical properties of GXLs are readily tuned by pressure. In most catalyzed reactions, it is imperative to recover and recycle the expensive catalyst. OATS provide a simple and efficient method for separating hydrophilic catalysts from organophilic substrates. They offer increased reaction rates and improved yields and selectivity – often seen in homogeneous systems – as well as simple, efficient separation (characteristic of heterogeneous separation) and recycle of the catalyst. As a demonstration, we conducted hydroformylations of hydrophobic 1-octene and *p*-methylstyrene in THF/H_2O and ACN/H_2O, respectively, under syngas pressure (mole ratio of 1:1 CO:H_2) to produce the corresponding branched and linear aldehydes in the presence of rhodium catalyst (Rh) with triphenylphosphine (TPP) ligands. Hydroformylation reactions are carried out at pressures of about 3 MPa (Blasucci *et al.* [31], El Ali *et al.* [32], Hallett *et al.* [33], Nair *et al.* [34]) and therefore; the use of CO_2 pressures (around 3.5 MPa) with the OATS does not require equipment modifications. Industrial hydroformylations are run in biphasic aqueous-organic systems, where the reaction occurs in the aqueous phase in the presence of the water-soluble sulfonated Rh/TPP catalyst and the products are extracted into the organic phase. As a consequence, this approach is practical only for water-soluble alkenes (mainly C3 and C4). OATS permits the hydroformylation of virtually insoluble alkenes such as 1-octene and p-methylstyrene to form the corresponding aldehydes. For example, we reported the hydroformylation of *p*-methylstyrene to produce 2-(*p*-tolyl)propanal (branched product) and 3-(*p*-tolyl)propanal in ACN/H_2O (70 vol% ACN) OATS (Figure 2). The branched aldehyde – the desired product – is a model for the intermediate (2-(4-isobutylphenyl)propanal) in the ibuprofen synthesis. The TOF was 92 at 40°C and 406 at 80°C and the yield of the branched product was 95% at 40°C and 80% at 80°C. The reduction in the yield of the branched product with temperature is attributed to β-hydride elimination,(Nair *et al.* [34]) where the intermediate complex of the catalyst with the branched product is converted back into the starting material at higher temperatures. The styrene conversion rates in OATS show at least an order of magnitude improvement over heterogeneously reported systems using solid supports and reactions run with ionic liquid modified silica sol-gel. Also, the branched product selectivity improves by more than 30%. The separation of the aldehyde is outstanding; at 2.5 MPa of CO_2 pressure, more than 99% of the branched aldehyde product partitions in the acetonitrile-rich phase.

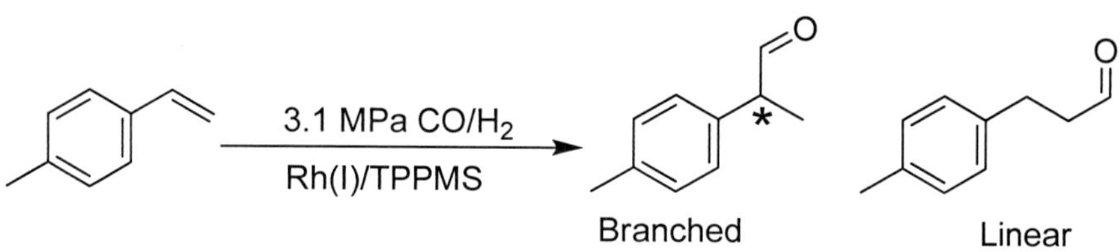

Figure 2: Hydroformylation of p-methylstyrene with Rh/TPPMS to produce branched (desired) and linear aldehydes.

2.2 Near critical water

Near critical water (NCW) is hot liquid water (critical point: 374ºC, 22.1MPa) at temperatures between 200-350°C. At near critical conditions, water has physical properties similar to acetone. At 275ºC, the density drops to 0.7-0.8 and the dielectric constant decreases to 20 at 275°C. This is mostly due to diminished hydrogen bonding. The solubility of nonpolar organics is markedly improved in NCW while it remains an effective medium for dissolving ionic compounds. Thus, it becomes possible to conduct reactions between hydrophobic organic molecules and hydrophilic ionic reactions under homogeneous conditions at the near critical temperature range and then, subsequently, easily isolate the organic product at room temperature upon phase split. Additionally, the dissociation constant of water K_W increases gradually from 10^{-14} at 25°C to about 10^{-11} at 275°C. This increase in K_W offers opportunities for acid or base catalysis, avoiding the need for added acids and bases and the subsequent waste-intensive neutralization steps.

We reported the Lewis acid-free Friedel-Crafts alkylation of phenol with *tert*-butyl alcohol in water at 275°C, as shown in Figure 3 (Chandler *et al.* [11]). Traditionally, these reactions use aluminum chloride ($AlCl_3$), which is corrosive, water-sensitive, difficult to handle, and often used in greater than stoichiometric quantities – leading to substantial waste generation. The yield of 2-*tert*-butylphenol reached 17 mol% after 30 hours of reaction time and then decreases to an equilibrium concentration of 10% and the yield of 4-*tert*-butylphenol increased linearly to 20% after 50 hours of reaction time. The 2-*tert*-butylphenol and 4-*tert*-butylphenol yields in NCW are similar to acid-catalyzed reactions reported in the literature with the use of five times excess of $HClO_4$. NCW plays a dual role as solvent and catalyst. Thus, NCW for Friedel-Crafts acylations could eliminate the use of mineral or Lewis acids and eliminate the subsequent neutralization steps that generate large quantities of waste salt for each pound of products.

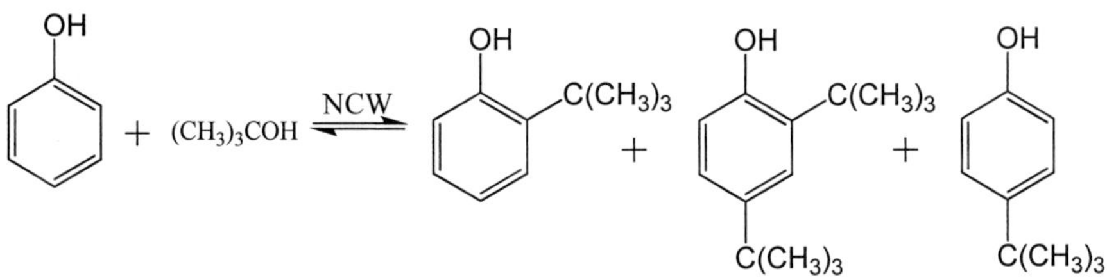

Figure 3: Alkylation of phenol with *tert*-butanol to produce 2-*tert*-butylphenol, 2,4-di-*tert*-butylphenol, and 4-*tert*-butylphenol.

3 Switchable solvents

3.1 Piperylene sulfone

Piperylene sulfone is synthesized from *trans*-1,3-pentadiene (*trans*-piperylene) and sulfur dioxide in the presence of a radical inhibitor (such as N-phenyl-2-naphtylamine). The synthesis can also be carried out using the commercially available and inexpensive 1,3-pentadiene, which is a mixture of trans- and cis-isomers. It should be noted that the cis-isomer reacts very slowly, if at all, with SO_2 to form the cyclic sulfone. We measured and reported the solvatochromic properties of piperylene sulfone, showing that piperylene sulfone has comparable solvent properties to dimethyl sulfoxide (DMSO, Vinci *et al.* [35]). The important difference between piperylene sulfone and DMSO is that the former is switchable and the latter is not. Indeed, at temperatures greater than 100ºC piperylene sulfone <u>*cleanly and efficiently*</u> decomposes to gaseous *trans*-1,3-pentadiene (b.p. 42ºC) and sulfur dioxide (b.p. –10ºC) via a retro-cheletropic process (Figure 4). The products can be collected and allowed to react to reform the piperylene sulfone solvent.

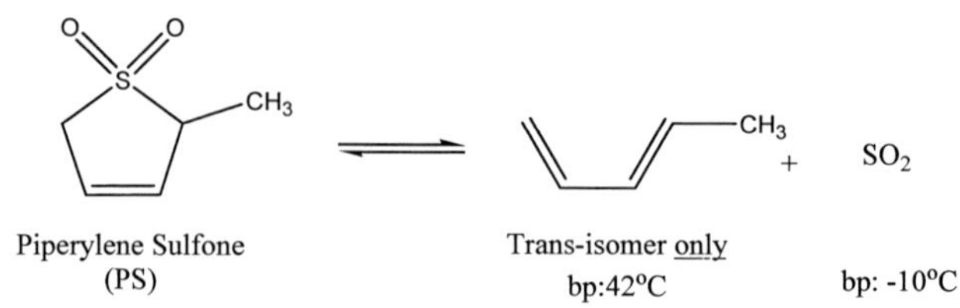

Figure 4: Reversible decomposition of piperylene sulfone into *trans*-piperylene and sulfur dioxide.

Critical to employing piperylene sulfone as a solvent for coupling reactions and separations is the requirement that decomposition of the solvent occurs with 100% efficiency. Differential scanning calorimetry coupled with thermogravimetric analysis was performed on piperylene sulfone. Heating at 5°C/min for 20 minutes and then holding at 120°C for 30 minutes triggered the decomposition, leaving no residual mass. Thus, in principle, facile product recovery and solvent recycle can be realized. Piperylene sulfone is one of a number of possible β,γ-unsaturated cyclic sulfones (also called sulfolenes) that can be employed as solvents. Others sulfolenes such as butadiene sulfone and isoprene sulfone are also of interest for switchable solvents applications. We reported the rates of substitution reaction of benzyl chloride with various anionic nucleophiles (Figure 5) in both piperylene sulfone and DMSO at 40ºC (Vinci *et al.* [35]). The rate constants are slightly greater in PS than in DMSO for the reaction of benzyl chloride with KSCN with values of 2.1±0.1 and 1.4±0.1 L/mol.s, respectively. The rate constants were similar for the reaction with potassium thioacetate and pyrrolidine-di-thiocarbamate. The addition of small amounts of water to the reaction mixtures improves the rate in both PS and DMSO for the reaction of benzyl chloride with KOAc, KCN, and KSCN. This improvement could result from the improved solubility of the reacting salts. After reaction, the mixtures were heated to 110ºC to decompose PS to gaseous *trans*-piperylene and sulfur dioxide. The solvent-free benzyl thiocyanate residue was isolated with 96% yield. The gases were collected and PS was reformed with 87% yield. The small size of the experimental set up is believed to be the reason for the loss of PS; greater solvent recovery yields are expected on larger scale.

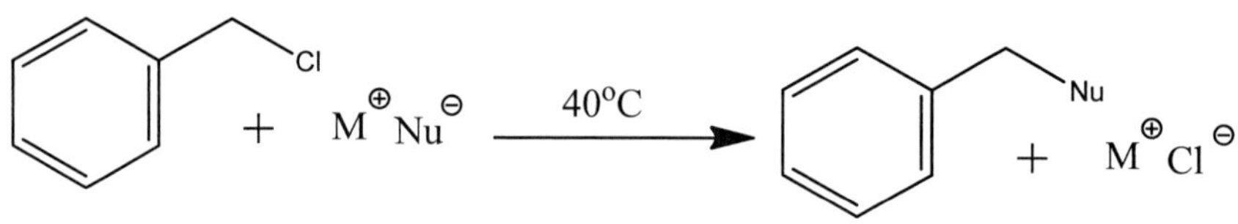

Figure 5: Nucleophilic displacement reactions.

3.2 Reversible ionic liquids (RevILs)

A reversible ionic liquid is a solvent system which can switch back and forth between a medium which is ionic and a medium which is non-ionic; the latter is referred to as a molecular medium. There are two major classes of reversible ionic liquids (RevILs): two-component and one-component. Two-component RevILs are based on an equimolar mixture of a neutral alcohol and a neutral molecule containing at least one basic nitrogen functionality. Examples of the latter are 1,8-diazabicyclo[5.4.0]undec-7-ene (DBU) and 2-butyl-1,1,3,3-tetramethylguanidine (TMBG, Phan *et al.* [36], Jessop *et al.* [37], Heldebrant *et al.* [38]). One-component systems eliminate the need for an alcohol as seen in the two-component RevILs; these systems employ solely a neutral molecule containing at least one basic nitrogen functionality [12]. An example of a one-component system are the silylated amines such as trialkoxy- or trialkyl-silylpropylamines as molecular liquid. Upon reaction with CO_2, these amine precursors form ionic liquids composed of the corresponding carbamate anion and ammonium cation pairs (Figure 6). The Lewis acid character of the silicon substitutent is believed to play a role in controlling the temperatures at which carbon dioxide is captured and released.

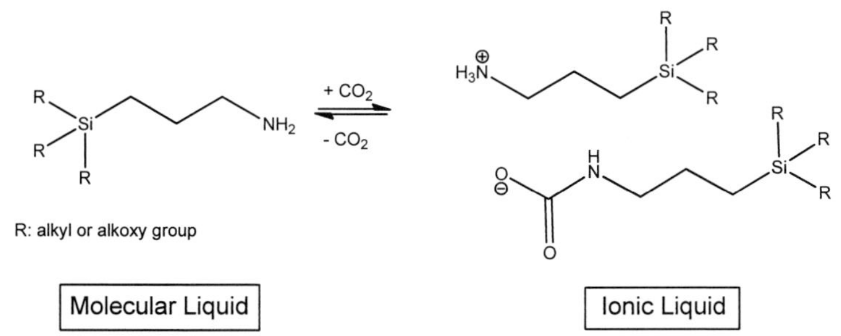

Figure 6: One-component system: reversible switch from a molecular liquid trialkoxy and trialkylsilylpropylamine to its corresponding ionic liquid upon addition of CO_2.

3.3 The Claisen-Schmidt condensation

The Claisen-Schmidt condensation of butanone and benzaldehyde in RevIL system (TMBG/MeOH) was reported by Hart *et al.* [39] (Figure 7). Three products are formed upon reaction: the internal enone (3-methyl-4-phenyl-but-3-en-2-one), the terminal enone (1-phenyl-pent-1-en-3-one) and water. Under basic conditions, the terminal enone product is the predominant product formed. It should noted that TMBG plays the dual role of base catalyst and solvent.

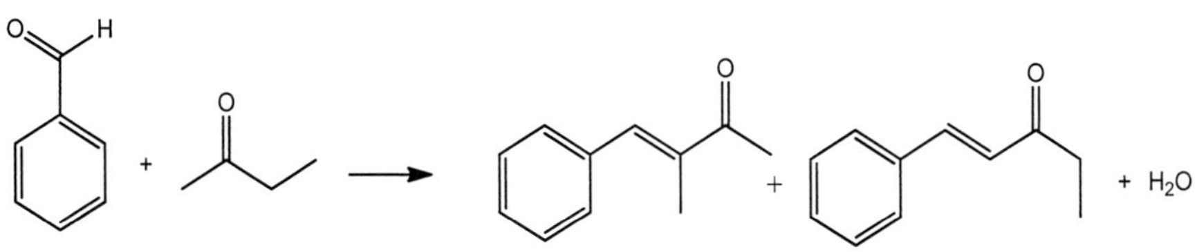

Figure 7: Claisen-Schmidt condensation of 2-butanone and benzaldehyde.

After 24 hours at room temperature or 3 hours at 80°C, yields of 48% and 44% in enone products were obtained, respectively. The formation of the terminal and internal enone-products was also studied as a function of time at 80°C. As the reaction time increased, the yields in enone products first increased from 13% at 1 hour to 44% at 3 hours and then began to decrease after 4 hours. The decrease was attributed to competing condensation processes between the enone products and the benzaldehyde, resulting in lower overall yields of the desired products. As a consequence, shorter reaction times to partial conversions were necessary in order to develop a process in which isolated yields were maximized and solvent recycle was possible. The isolation of the enone products was performed by adding n-octane and methanol to the reaction mixture, followed by the addition of CO_2, which triggered the formation of the ionic liquid. Under these conditions an octane phase separated from the newly developed ionic liquid phase. The enone products were predominantly soluble in the octane phase and were easily separated by decantation.

3.4 CO_2 capture

RevILs are also being studied for absorbing CO_2 from various flue gas streams, where they may require less energy for CO_2 capture and release than the conventional alkanolamine process. The presence of large volumes of water (commonly 70 vol%) increases the energy demand for the thermal regeneration of the aqueous alkanolamines. RevILs behave in a dual capacity as solvents for CO_2 capture: (1) the chemisorption of CO_2 upon reaction of the molecular liquid to form RevIL and (2) by the physical absorption of the CO_2. We reported the use of trimethoxy (TMSA), triethoxy (TESA), triethyl (TEtSA), and tripropyl (TPSA) silated propylamines for CO_2 capture (Blasucci *et al.* [40, 41]). The chemical capacities can be enhanced by the physisorption capabilities of the RevILs. The physisorption is a much less exothermic process, thus the regeneration of physisorbed CO_2 requires less energy.

4 Conclusions

Solvents are necessary for most chemical processes and lead to a large percentage of the process cost and to the magnitude of the waste-stream. In order to achieve chemical processes that are both competitive and environmentally conscious, solvents must address simultaneously reaction efficiency, product separation, and recyclability. Tunable solvents and smart switchable solvents can improve specific families of chemical reactions and processes. These solvents change properties upon application of physical or chemical stimuli and, as a consequence, provide facile separation techniques. Smart switchable solvents change properties abruptly and reversibly. Unlike DMSO, because of a facile and reversible reaction between trans-piperylene and sulfur dioxide, piperylene sulfone allows for easy product isolation and solvent recycle are easily accomplished. In addition, Claisen-Schmidt condensation was successfully carried out in RevILs with the added capabilities of improved separation of the product and recycling of the solvent system. Reversible ionic liquids facilitate catalyst and product recovery, yielding efficient separations of catalysts for reuse. Each of these different solvents has advantages and limitations. The feasibility of their applications depend on the specifics of each reaction or process. Importantly, it should be recognised that solvents can accomplish more than heat and mass transfer; they can actively contribute to facilitate reaction and product separation while minimizing waste generation.

References

[1] Dillow, A.K., Brown, J.S., Liotta, C.L., Eckert, C.A., Supercritical fluid tuning of reactions rates: The cis-trans isomerization of 4-4 '-disubstituted azobenzenes, *J Phys Chem A*, 102, 7609-7617, 1998.

[2] Brown, J.S., Lesutis, H.P., Lamb, D.R., Bush, D., Chandler, K., West, B.L., Liotta, C.L., Eckert, C.A., Schiraldi, D., Hurley, J.S., Supercritical fluid separation for selective quaternary ammonium salt promoted esterification of terephthalic acid, *Ind Eng Chem Res*, 38, 3622-3627, 1999.

[3] Thompson, R.L., Glaser, R., Bush, D., Liotta, C.L., Eckert, C.A., Rate variations of a hetero-Diels-Alder reaction in supercritical fluid CO2, *Ind Eng Chem Res*, 38, 4220-4225, 1999.

[4] Brown, R.A., Pollet, P., McKoon, E., Eckert, C.A., Liotta, C.L., Jessop, P.G., Asymmetric hydrogenation and catalyst recycling using ionic liquid and supercritical carbon dioxide, *J Am Chem Soc*, 123, 1254-1255, 2001.

[5] Nolen, S.A., Lu, J., Brown, J.S., Pollet, P., Eason, B.C., Griffith, K.N., Glaser, R., Bush, D., Lamb, D.R., Liotta, C.L., Eckert, C.A., Thiele, G.F., Bartels, K.A., Olefin epoxidations using supercritical carbon dioxide and hydrogen peroxide without added metallic catalysts or peroxy acids, *Ind Eng Chem Res*, 41, 316-323, 2002.

[6] Eckert, C.A., Liotta, C.L., Bush, D., Brown, J.S., Hallett, J.P., Sustainable reactions in tunable solvents, *J Phys Chem B*, 108, 18108-18118, 2004.

[7] Koch, D., Leitner, W., Rhodium-catalyzed hydroformylation in supercritical carbon dioxide, *J Am Chem Soc*, 120, 13398-13404, 1998.

[8] Furstner, A., Ackermann, L., Beck, K., Hori, H., Koch, D., Langemann, K., Liebl, M., Six, C., Leitner, W., Olefin metathesis in supercritical carbon dioxide, *J Am Chem Soc*, 123, 9000-9006, 2001.

[9] Solinas, M., Jiang, J.Y., Stelzer, O., Leitner, W., A cartridge system for organometallic catalysis: Sequential catalysis and separation using supercritical carbon dioxide to switch phases, *Angew Chem Int Edit*, 44, 2291-2295, 2005.

[10] Maayan, G., Ganchegui, B., Leitner, W., Neumann, R., Selective aerobic oxidation in supercritical carbon dioxide catalyzed by the H5PV2Mo10O40 polyoxometalate, *Chem Commun*, 2230-2232, 2006.

[11] Chandler, K., Deng, F.H., Dillow, A.K., Liotta, C.L., Eckert, C.A., Alkylation reactions in near-critical water in the absence of acid catalysts, *Ind Eng Chem Res*, 36, 5175-5179, 1997.

[12] Chandler, K., Liotta, C.L., Eckert, C.A., Schiraldi, D., Tuning alkylation reactions with temperature in near-critical water, *Aiche J*, 44, 2080-2087, 1998.

[13] Lesutis, H.P., Glaser, R., Liotta, C.L., Eckert, C.A., Acid/base-catalyzed ester hydrolysis in near-critical water, *Chem Commun*, 2063-2064, 1999.

[14] Patrick, H.R., Griffith, K., Liotta, C.L., Eckert, C.A., Glaser, R., Near-critical water: A benign medium for catalytic reactions, *Ind Eng Chem Res*, 40, 6063-6067, 2001.

[15] Nolen, S.A., Liotta, C.L., Eckert, C.A., Glaser, R., The catalytic opportunities of near-critical water: a benign medium for conventionally acid and base catalyzed condensations for organic synthesis, *Green Chem*, 5, 663-669, 2003.

[16] Jessop, P.G., Subramaniam, B., Gas-expanded liquids, *Chem Rev*, 107, 2666-2694, 2007.

[17] Ablan, C.D., Hallett, J.P., West, K.N., Jones, R.S., Eckert, C.A., Liotta, C.L., Jessop, P.G., Use and recovery of a homogeneous catalyst with carbon dioxide as a solubility switch, *Chem Commun*, 2972-2973, 2003.

[18] Jessop, P.G., Stanley, R.R., Brown, R.A., Eckert, C.A., Liotta, C.L., Ngo, T.T., Pollet, P., Neoteric solvents for asymmetric hydrogenation: supercritical fluids, ionic liquids, and expanded ionic liquids, *Green Chem*, 5, 123-128, 2003.

[19] Daintree, L.S., Kordikowski, A., York, P., Separation processes for organic molecules using SCF technologies, *Advanced Drug Delivery Reviews*, 60, 351-372, 2008.

[20] Eckert, C.A., Bush, D., Brown, J.S., Liotta, C.L., Tuning solvents for sustainable technology, *Industrial & Engineering Chemistry Research*, 39, 4615-4621, 2000.

[21] Eckert, C.A., Chandler, K., Tuning fluid solvents for chemical reactions, *J Supercrit Fluid*, 13, 187-195, 1998.

[22] Eckert, C.A., Knutson, B.L., Debenedetti, P.G., Supercritical fluids as solvents for chemical and materials processing, *Nature*, 383, 313-318, 1996.

[23] Eckert, C.A., Vanalsten, J.G., Stoicos, T., Supercritical Fluid Processing, *Environ Sci Technol*, 20, 319-325, 1986.

[24] Jessop, P.G., Homogeneous catalysis using supercritical fluids: Recent trends and systems studied, *Journal of Supercritical Fluids*, 38, 211-231, 2006.

[25] Ramsey, E., Sun, Q.B., Zhang, Z.Q., Zhang, C.M., Gou, W., Mini-Review: Green sustainable processes using supercritical fluid carbon dioxide, *Journal of Environmental Sciences-China*, 21, 720-726, 2009.

[26] Rayner, C.M., The potential of carbon dioxide in synthetic organic chemistry, *Organic Process Research & Development*, 11, 121-132, 2007.

[27] Rezaei, K., Temelli, F., Jenab, E., Effects of pressure and temperature on enzymatic reactions in supercritical fluids, *Biotechnology Advances*, 25, 272-280, 2007.

[28] Tang, Z., Xie, W.H., Zong, B.N., Min, E.Z., Recent research progress of chemical reactions under supercritical conditions, *Chinese Journal of Chemical Engineering*, 12, 498-504, 2004.
[29] Hines, A.L., Maddox, R.N., Mass Transfer Fundamentals and Applications Prentice Hall, New Jersey, 1985.
[30] Bird, R.B., Stewart, W.E., Lightfoot, E.N., Transport Phenomena, Second ed., Wiley India, New Delhi, 2006.
[31] Blasucci, V.M., Husain, Z.A., Fadhel, A.Z., Donaldson, M.E., Vyhmeister, E., Pollet, P., Liotta, C.L., Eckert, C.A., Combining Homogeneous Catalysis with Heterogeneous Separation using Tunable Solvent Systems, *Journal of Physical Chemistry A*, 114, 3932-3938, 2010.
[32] El Ali, B., Tijani, J., Fettouhi, M., El-Faer, M., Al-Arfaj, A., Rhodium(I) and rhodium(III)-heteropolyacids supported on MCM-41 for the catalytic hydroformylation of styrene derivatives, *Appl Catal a-Gen*, 283, 185-196, 2005.
[33] Hallett, J.P., Ford, J.W., Jones, R.S., Pollet, P., Thomas, C.A., Liotta, C.L., Eckert, C.A., Hydroformylation catalyst recycle with gas-expanded liquids, *Industrial & Engineering Chemistry Research*, 47, 2585-2589, 2008.
[34] Nair, V.S., Mathew, S.P., Chaudhari, R.V., Kinetics of hydroformylation of styrene using homogeneous rhodium complex catalyst, *Journal of Molecular Catalysis a-Chemical*, 143, 99-110, 1999.
[35] Vinci, D., Donaldson, M., Hallett, J.P., John, E.A., Pollet, P., Thomas, C.A., Grilly, J.D., Jessop, P.G., Liotta, C.L., Eckert, C.A., Piperylene sulfone: a labile and recyclable DMSO substitute, *Chem Commun*, 1427-1429, 2007.
[36] Phan, L., Chiu, D., Heldebrant, D.J., Huttenhower, H., John, E., Li, X.W., Pollet, P., Wang, R.Y., Eckert, C.A., Liotta, C.L., Jessop, P.G., Switchable solvents consisting of amidine/alcohol or guanidine/alcohol mixtures, *Ind Eng Chem Res*, 47, 539-545, 2008.
[37] Jessop, P.G., Heldebrant, D.J., Li, X.W., Eckert, C.A., Liotta, C.L., Green chemistry - Reversible nonpolar-to-polar solvent, *Nature*, 436, 1102-1102, 2005.
[38] Heldebrant, D.J., Jessop, P.G., Thomas, C.A., Eckert, C.A., Liotta, C.L., The reaction of 1,8-diazabicyclo[5.4.0]undec-7-ene (DBU) with carbon dioxide, *J Org Chem*, 70, 5335-5338, 2005.
[39] Hart, R., Pollet, P., Hahne, D.J., John, E., Llopis-Mestre, V., Blasucci, V., Huttenhower, H., Leitner, W., Eckert, C.A., Liotta, C.L., Benign coupling of reactions and separations with reversible ionic liquids, *Tetrahedron*, 66, 1082-1090, 2010.
[40] Blasucci, V., Hart, R., Mestre, V.L., Hahne, D.J., Burlager, M., Huttenhower, H., Thio, B.J.R., Pollet, P., Liotta, C.L., Eckert, C.A., Single component, reversible ionic liquids for energy applications, *Fuel*, 89, 1315-1319, 2010.
[41] Blasucci, V.M., Hart, R., Pollet, P., Liotta, C.L., Eckert, C.A., Reversible ionic liquids designed for facile separations, *Fluid Phase Equilibria*, 294, 1-6, 2010.

A new and greener method to manufacture copolymer-1

E. Ponnusamy
Sigma-Aldrich, USA

Abstract

Polyamino acid random copolymers have a wide variety of properties that mimic proteins, which make these random copolymers suitable for the treatment of certain diseases. For example, polyamino acid random copolymers comprising alanine, glutamic acid, lysine, and tyrosine (copolymer-1) have been used in the treatment of multiple sclerosis. Synthesis of this copolymer requires protecting groups for the side chains of the glutamic acid and lysine residues. The manufacture of copolymer-1 is typically achieved by protecting lysine with N^{ε}-trifluoroacetyl (TFA) and glutamic acid with γ-benzyl protecting groups respectively. Deprotecting TFA and benzyl protecting groups require two separate steps. Removal of the benzyl-protecting group from glutamic acid requires using hazardous chemicals (HBr/acetic acid) and highly flammable solvents and also generates a by-product, benzyl bromide a very strong lachrymator. Removal of the TFA group from lysine, the product needs to be further treated with flammable piperidine. These deprotection methods require long reaction times and handling, which in-turn reduces yield. Therefore, a need exists for an efficient, greener and less toxic process to manufacture copolymer-1. A new greener method for the synthesis of Copolymer-1 has been developed. The method is greener, efficient, and uses less toxic reagents than literature methods. The present invention provides a method for the synthesis of copolymer-1 by polymerizing a mixture of N-carboxyanhydrides of alanine, tyrosine, γ-ethyl-protected-glutamic acid, and N^{ε}-TFA-protected-lysine in the presence of a polymerization initiator to form a protected polyamino acid random copolymer. Then adding a base to the protected polyamino acid random copolymer to cleave both the ethyl group from the glutamic acid and the TFA group from the lysine in a single deprotection step and also eliminated a large volume of hazardous chemical usage and waste generation. The new method

WIT Transactions on Ecology and the Environment, Vol 154, © 2011 WIT Press
www.witpress.com, ISSN 1743-3541 (on-line)
doi:10.2495/CHEM110041

increased copolymer-1 yield, scalability, and quality and also reduced the cycle time.

Keywords: polyamino acid copolymers, protecting group, deprotection, dialysis, lyophilisation.

1 Introduction

Multiple sclerosis is a chronic inflammatory demyelinating disease of the central nervous system, in which infiltrating lymphocytes, predominantly T-cells and macrophages, cause damage of the myelin sheath. Copolymer-1 has been shown to alter positively the natural history of multiple sclerosis by both reducing the relapse rate and affecting disability [1–4]. It has been known to have suppressive and protective effects in experimental autoimmune encephalomyelitis, which can be induced in different species by various encephalitogenic antigens [5–9]. Copolymer-1, a synthetic polymer of amino acids, has a specific effect on the autoimmune process involved in experimental allergic encephalomyelitis and multiple sclerosis [10]. A major goal in the treatment of autoimmune disease has been the development of antigen-specific immunomodulating therapies [11].

Copolymer-1 is a randomized mixture of synthetic polypeptides consisting of the four naturally occurring amino acids L-alanine, L-glutamic acid, L-lysine and L-tyrosine. The techniques used for the polymerization of the N-carboxyanhydrides (NCA) to polymers are known to those skilled in the art and are given in detail in the review article by Goodman and Peggion [12], and the book by Kricheldorf [13] and also the recent publications by van Dijk-Wolthuis et al. [14]. The following synthesis specifies greener synthesis of poly (L-Ala, L-Glu, L-Lys, L-Tyr) using N-carboxyanhydrides of L-alanine, γ-ethyl-L-glutamic acid, N^{ε}-TFA-L-lysine and L-tyrosine.

2 Experimental

All materials used in this study were from Sigma-Aldrich. Varian Inova AS400 Nuclear Magnetic Resonance spectrometer (NMR) was used for NMR analysis. Dawn EOS by Wyatt Technologies was used for light scattering analysis and Cannon-Fenske viscometer, size 50 was used for viscosity measurements. Amino acid analysis was done using Pickering PCX5200 with Hitachi L7400 detector.

2.1 Greener synthesis of Poly (L-Ala, L-Glu, L-Lys, L-Tyr) Acetate (Copolymer-1)

2.1.1 Polymerization

Poly (L-Ala, γ-Et-L-Glu, N^{ε}-TFA-L-Lys, L-Tyr) is prepared as follows: 3.334g (0.029 mole) of L-alanine NCA, 1.942g (0.010 mole) of γ-ethyl-L-glutamic acid NCA, 6.479g (0.024 mole) of N^{ε}-TFA-L-lysine NCA and 1.000g (0.005 mole) of L-tyrosine NCA are dissolved in 0.225 liter of 1,4-dioxane to make a 0.3M solution. Add ~1.25g of charcoal and filter to get clear colorless solution. The filtered NCA solution is transferred to a 1 liter three neck RB flask equipped with mechanical mixing and a water bath at a temperature of 25–30°C. 2.7 ml of 1N sodium methoxide (0.0027 moles) is placed in 25 ml of 1,4-dioxane. The sodium methoxide solution is added to the NCA solution in one portion with vigorous mixing. Polymerization proceeded at room temperature (25–30°C) with evolution of carbon dioxide and the polymer solution is mixed for 2 hours and held at 25–30°C for 18 - 24 hours.

2.1.2 Precipitation of protected polymer

Slowly pour the polymer solution in 1,250ml of DI-water with vigorous mixing. Protected polymer precipitates out and mix for 30 minutes. The product is filtered and washed the polymer with DI-water.

2.1.3 Deprotection of ethyl and TFA groups

Transfer the filtered polymer to a 500ml Erlenmeyer flask. Add 140ml of 1.5M ethanolic sodium hydroxide and mix for 0.5 hour. Polymer goes into solution within 5 minutes of adding the reagent. Dilute the polymer solution with ~150ml of DI-water and mix for 5 minutes.

2.1.4 Dialysis and lyophilization (freeze drying)

Dialyze the deprotected polymer solution (18–24 hours) against running deionized water using ~12K molecular weight cut-off dialysis tubings. This process removes the oligomers, low molecular weight polymers and salts. The dialysis tubes are transferred to 3.5% acetic acid solution (~18L) and let stand for 7 hours and slowly mix the solution containing the dialysis tubings to make salt exchanges. Then dialyze the solution against running de-ionized water for 18–24 hours to remove the excess acetic acid, collect the solution, filter through a 0.2μm filter and lyophilize (freeze dry) to get the solid poly (L-Ala, L-Glu, L-Lys, L-Tyr) acetate (copolymer-1). Yield: 6.21g.

3 Results and discussions

^{1}H and ^{13}C NMRs showed the complete removal of the ethyl and TFA groups and also consistent with structure (Figures 1, 2). Measured specific viscosity in

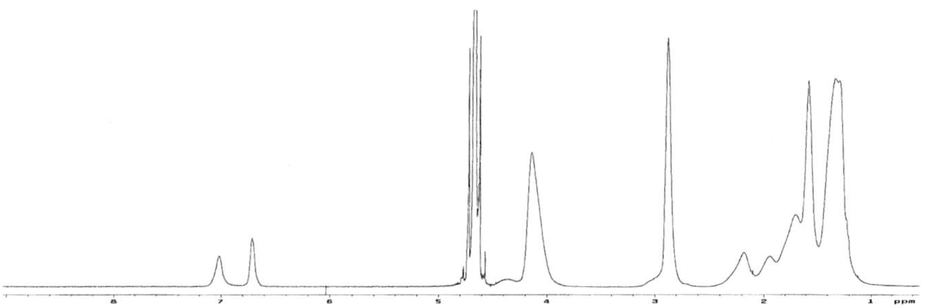

Figure 1: ^{1}H NMR of copolymer-1.

water, 1% solution at 25ºC and calculated the viscosity molecular weight [15].

Also analyzed the polymer by light scattering (GPC MALLS) and determined molecular weight (Mw) (Figure 3). Molecular weight profile and polydispersity are similar to that of the old method. Measured specific rotation in water (1% solution) at 25ºC and the results are comparable. Amino acid ratios are determined by amino acid analysis and are within the expected range (Table 1).

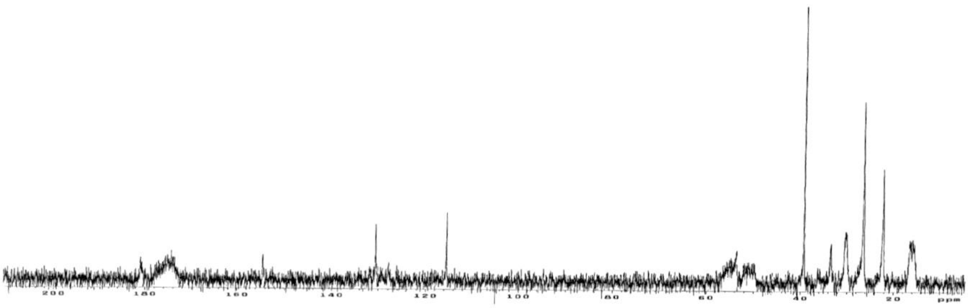

Figure 2: ^{13}C NMR of copolymer-1.

Table 1: Molecular Weights, Polydispersity, Specific Rotation, Amino Acid Analysis and Yield for Copolymer-1.

Viscosity Molecular Weight	GPC MALLS Molecular Weight	Polydispersity (Mw/Mn)	Specific Rotation	Amino Acid Analysis	% Yield
18,000	21,300	1.2	-63.2°	Ala: 5.70 Glu: 1.90 Lys: 4.80 Tyr: 1.00	71.9

Figure 3 shows the GPC MALLS profile for copolymer-1 and is consistent with the expected molecular weight and polydispersity.

The main purpose of the synthesis of copolymer-1 described above was to eliminate the usage of hazardous reagents (HBr/acetic acid and piperidine) and make the process more greener and less labor intensive. Aqueous alkaline condition is used in the deprotection step of Poly (L-Ala, γ-Et-L-Glu, N^{ε}-TFA-L-Lys, L-Tyr) copolymer to remove the TFA group from lysine and the ethyl group from glutamic acid in a single reaction. No racemization occurs during the removal of these protecting groups, as is apparent from a comparison of the specific rotation of copolymer-1 prepared via benzyl protected glutamic acid (-65° ± 5°) with that of ethyl protected (Table 1).

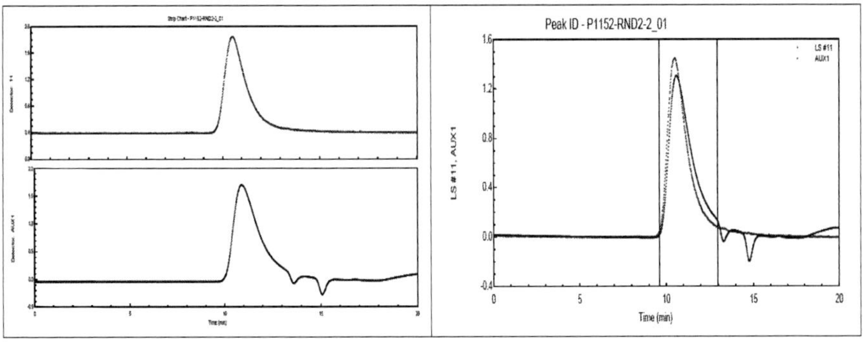

Figure 3: GPC MALS analysis instrument type: DAWN EOS; Cell type: K5; Laser wavelength: 690.0 nm; Solvent: 20mM Sodium Phosphate +130mM Sodium Chloride.

4 Conclusions

General procedure to make copolymer-1 was to polymerize N-carboxyanhydrides of L-alanine, γ-benzyl-L-glutamic acid, N^{ε}-TFA-L-lysine and L-tyrosine to make protected copolymer. Then remove benzyl group from glutamic acid with HBr/acetic acid and TFA group from lysine with piperidine in multiple steps and also a large volume of hazardous wastes have been generated. In the new process, eliminated the usage of hazardous HBr/acetic acid and piperidine reagents by replacing benzyl with ethyl protecting group for glutamic acid followed by a simple one step base deprotection. Also the new process eliminated large volume of hazardous waste generation. The new greener process decreased the cycle time to half, decreased the manufacturing cost, provided an unlimited scalability potential and also increased the yield without compromising quality.

References

[1] Johnson K. P., J. Neurol., **243**, (Suppl. 1), S3-S7 (1996).
[2] Teitelbaum D., Milo R., Arnon R. and Sela M., Proc. Natl. Acad. Sci. USA, **89**, 137-141 (1992).
[3] Comi G. and Moiola L., Baillieres Clin. Neurol. Oct. 6 (3), 495-509 (1997).
[4] Teitelbaum D., Arnon R. and Sela M., Cell Mol. Life Sci. **53**, 24-28 (1997).
[5] Ziemssen T., Kumpfel T., Klinkert, W.E.F., Neuhaus, O. and Hohlfeld, R., Brain **125**, 2381-2391 (2002).
[6] Teitelbaum D., Webb C., Meshorer A., Arnon R. and Sela M., Nature **240**, 564-566 (1972).
[7] Web C., Teitelbaum D., Abramsky O., Arnon R. and Sela M., Isr. J. Med. Sci. **11**, 1388 (1975).

[8] Teitelbaum D., Fridkis-Hareli M., Arnon R. and Sela M., J. Neuroimmunol. **64**, 209-217 (1996).
[9] Sela M., Int. Rev. Immunol. **18**, 201-216 (1999).
[10] Sela M., Polym. Adv. Technol. **13**, 859-862 (2002).
[11] Fridkis-Hareli M., Santambrogio L., Stern J.N.H., Fugger L., Brosnan C. and Strominger J.L., J. Clin. Invest. **109**, 1635-1643 (2002).
[12] Goodman M. and Peggion E., Pure and Applied Chemistry, **53**, 699-714 (1981).
[13] Kricheldorf H. R., "Alpha Amino Acids-N-Carboxyanhydrides and Related Heterocycles", Springer Verlag (1987).
[14] van Dijk-Wolthuis W. N. E., van de Water L., van de Wetering P., van Steenbergen M. J., Kettenes-van den Bosch J. J., Schuyl W. J. W. and Hennink W. E., Macromol. Chem. Phys. **198**, 3893-3906 (1997).
[15] Yaron A. and Berger A., Biochim. Biophys. Acta **69**, 397-399 (1963).

Corning® Advanced-Flow™ reactor technology for process intensification

D. Chamrai
Corning SNG, Corning Scientific Center, St Petersburg, Russia

Abstract

The concept of process intensification has been around for at least a decade in chemical process engineering. Conventional "batch" synthesis, with mixing and reactions done in bulky meter-scale vessels, often generates by-products and suffers from energy and material waste. Also, safety could become a concern when high-energy processes or highly toxic reagents are involved. As one of its key aspects, process intensification considers development of new equipment allowing substantially smaller, cleaner, safer and more energy-efficient and scalable methods for industrial-scale production of pharmaceutical and specialty chemical products. These needs can be addressed through the continuous-flow technology of Corning's Advanced-Flow™ glass reactors, which consist of modules with millimeter-scale channels that allow optimization of thermal and interfacial mass transfer for reducing overall heat and mass transfer resistances. This paper will describe Corning® Advanced-Flow™ reactor technology and several application cases in which these reactors demonstrated their high potential for industrial deployment and process intensification due to their ability to increase the efficiency, scalability, and quality of chemical processing – all while reducing environmental impact, performance variability and cost.
Keywords: continuous flow, advanced-flow reactor, micro-channel, process intensification, green chemistry.

1 Introduction

Increasing demand from recovering end markets and growing economies will require expansion of existing production capacities in specialty, fine, pharmaceuticals, agrochemicals and other chemical industry segments, again attracting attention to chemical processes intensification. Production of modern

WIT Transactions on Ecology and the Environment, Vol 154, © 2011 WIT Press
www.witpress.com, ISSN 1743-3541 (on-line)
doi:10.2495/CHEM110051

chemicals uses 8% of hydrocarbons worldwide - mostly as raw material, or "feedstock," but also for energy used in the manufacturing process [1]. According to DOE data the U.S. chemical industry accounts for almost 30% of all U.S. industrial energy consumption [2]. The EU chemical industry is also rather energy-intensive. Total consumption by chemical plants in EU (160Mtoe) accounts for roughly 3% of global and about 12% of EU energy demand [3]. In the chemicals industry, energy costs account for on average 10%-15% of manufacturing costs [4]. Simultaneously, changing carbon emissions laws around the world have made many chemical plants worried about their competitiveness, since the high costs associated with greener manufacturing processes cannot be passed on to consumers. Another aspect of the chemical industry's sustainability le intensification is focus is on minimization of environmental impact. Despite currently available waste treatment technologies, the problem of waste reduction is still a significant matter for many chemical plants. Thus, any technology allowing a decrease in energy consumption and minimization of waste generation may bring significant advantage for the chemicals manufacturers and contribute to higher sustainability of the chemical industry.

Several chemical industry segments, such as petrochemicals and polymers production have been utilizing continuous or semi-continuous flow production processes for a long time, benefiting from reduced process hold-ups, good residence time control, enhanced safety, high yields and selectivity and so on [5]. Today, new technologies are appearing, allowing segments of the chemical industry, such as pharmaceuticals and fine chemicals production, which traditionally employed predominantly batch process, to start switching to continuous production and thus make a step forward in chemical process evolution.

2 Corning® Advanced-Flow™ reactor

Corning's Advanced-Flow™ reactors represent one of the possible solutions which open a way for broader employment of continuous-flow production across chemical industry segments. The key principle of the technology is based on use of a continuous-flow modular system, which consists of a set of various fluidic glass modules, where reaction layer and heat exchange layers are combined in a single block, thus allowing excellent control over the reaction temperature due to elimination of hot spots. The internal space of the reactor layer is arranged as a set of continuously connected micro-channels with millimeter-scale dimensions, which allows reaching optimal surface-to-volume ratio of the reactor, (fig. 1.) Due to the small channel dimensions, such reactors are often referred as "microreactors". For multiphase reaction systems efficient mixing often becomes very critical. Geometry of channels of the Corning® Advanced-Flow™ reactor is designed in a way which allows excellent mixing at various phases [6]. Channel walls are made from materials with good resistance to chemical corrosion and smooth surfaces that enable easy cleaning. Scaling up from laboratory scale to production scale still remains an issue for many chemical reactions employed in

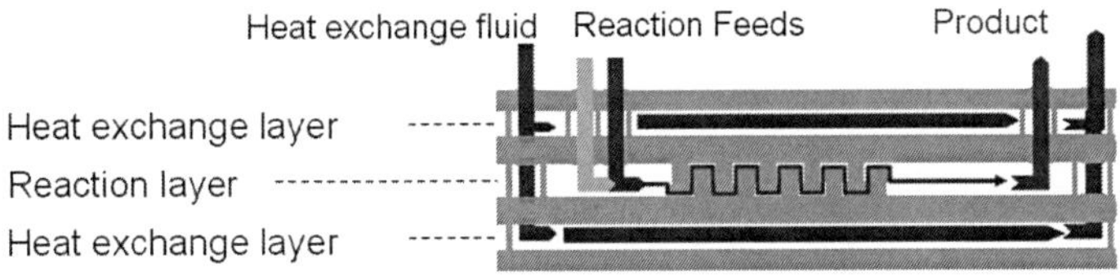

Figure 1: Construction and operation principle of a Corning® Advanced-Flow™ reactor fluidic module.

pharmaceuticals and fine products manufacturing. Corning® Advanced-Flow™ reactors deliver industrial scale capacity via both true scaling-out of the reactor volume and numbering-up principle.

Today, mass flow of a single Corning reactor on an annual basis varies between 7 and 1600 metric tons due to employment of various generations and scales of fluidic modules. Further capacity increases can be easily achieved by increase the number of reactors and their parallel use in production. Necessary to emphasize also a more general advanced feature of continuous flow processes such as an opportunity to engineer and optimize the equipment with respect to a particular reaction to be performed opposed to the traditional batch approach, when reaction is engineered in the most suitable way to fit available equipment.

3 Application examples of Corning® Advanced-Flow™ reactors

Tens of reactions have already been successfully tested with Corning® Advanced-Flow™ reactors, including reactions with miscible and immiscible liquids, gases utilization and gases release as well as reactions involving solids formation. Translating lab-scale experiments or pilot-scale tests to the industrial scale production line, we've made estimates of the economic advantage of Corning reactor-based lines over comparable traditional batch processes for several tested applications. Below, detail is provided on two examples of industrial scale operations assessment. For each case, the assessment of potential economic effect was made considering the following five process characteristics: utilities and raw materials costs, labour, capital depreciation cost, process yield and product quality as well as product throughput.

The first case covers a nitration plant with an annual capacity of 400 tons of end-product. The basis for this case is a pilot system being successfully used for selective nitration [7]. Significant benefits for this type of application can be obtained due to the ability to operate at higher concentrations, thus reducing use of solvent, while at the same time improving yield and selectivity. Other factors contributing to cost benefits include significant reduction of the size of the downstream equipment, smaller production line footprint and lower raw material inventory. Additionally, this test case demonstrated the ability to achieve significant reduction in the development cycle, moving from laboratory to production in less than 16 months. Comparison with the same using batch

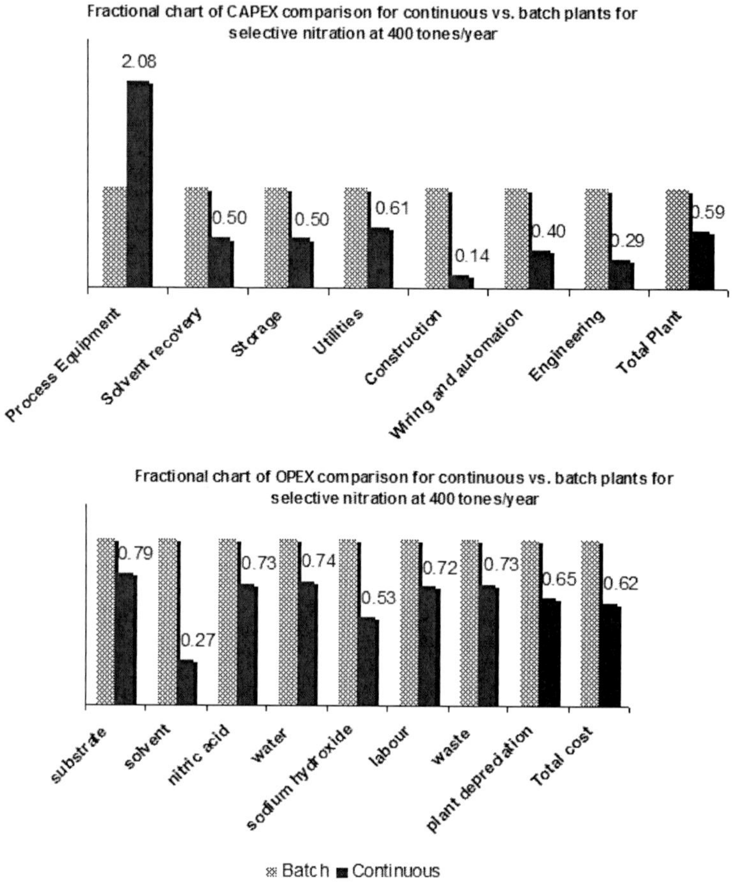

Figure 2: Fractional CAPEX and OPEX comparison of batch vs. continuous flow production plant for selective nitration.

process was made, illustrating achievable economic impacts on capital investment and operation cost, fig. 2.

Another example evaluation of Corning® Advanced-Flow™ reactor application is a hydrogenation plant with production capacity of 400 tones/year. The reaction performed is a selective highly exothermic hydrogenation, with application of a noble metal catalyst in slurry combined with hydrogen gas. This case demonstrates a different set of economic advantages of continuous-flow processes over batch production, since the optimized batch reaction demonstrated 100% yield and selectivity [8]. The advantage of continuous process in this case is related to drastic reduction of the reaction time – about 2 minutes for continuous vs. 10 hours in batch. Other benefits of continuous over batch in this case include reduction in the amount of expensive catalyst being used, smaller equipment footprint, and lower cost of the production facility due

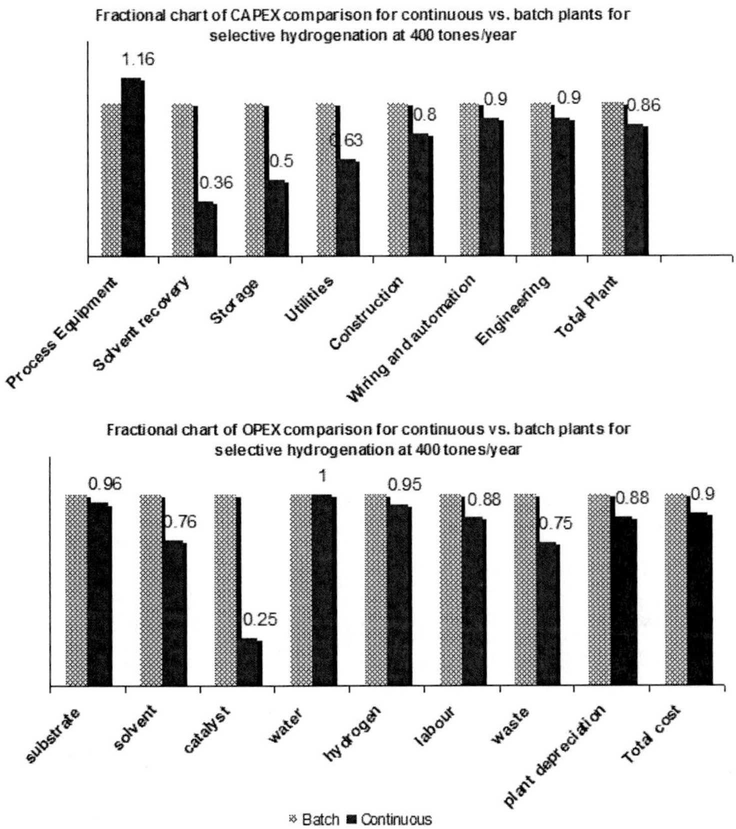

Figure 3: Fractional CAPEX and OPEX comparison of batch vs. continuous flow production plant for selective hydrogenation.

to a decrease in the level of safety requirements. The fractional cost comparisons for hydrogenation plant are presented below, fig. 3.

Significant additional advantages can also be achieved, especially in the case of a green-field plant, due to the much smaller plant footprint required. For instance, based on Corning's assessment, the batch line for nitration operating at capacity of 400 tones/year, will require a footprint of 2,927.52 m^2, whereas the same production line, employing Corning equipment can be allocated on an area of 455.84 m^2. Afterwards, maintenance and utility costs for a smaller production facility will continue to bring additional financial benefit and reduced energy consumption.

Of course, as any known technology Corning® Advanced-Flow™ reactor may have some limitations in applicability and economic benefits. Detailed analysis of applicability and estimates for potential economic benefits should be conducted on a case-by-case basis.

4 Efficient and green

It is clear that increasing the efficiency of manufacturing processes in the chemical industry could result in substantial benefit for the industry and increase its overall sustainability, lowering the negative impact on the environment. Unfortunately, the complexity of the thousands of processes being used in chemical manufacturing and significant differences between different plants' energy utilization patterns do not allow precise evaluation of the potential economic and environmental impact of specific process intensification efforts. However, in some cases applying a number of assumptions, we can try to assess potential impacts of specific technologies on the global environment. Averaging our results on the assessment of the potential benefits of Corning® Advanced-Flow™ reactors used in comparison with similar batch processes, the cumulative economic and environmental effects have been estimated and a graphical summary has been generated, fig. 4 [9]. The chart illustrates that significant improvements in process economics can be achieved due to substantial decreases in risks associated with process scale up. Also, significant reduction of solvent utilization makes a great contribution toward much greener and less expensive production of chemicals.

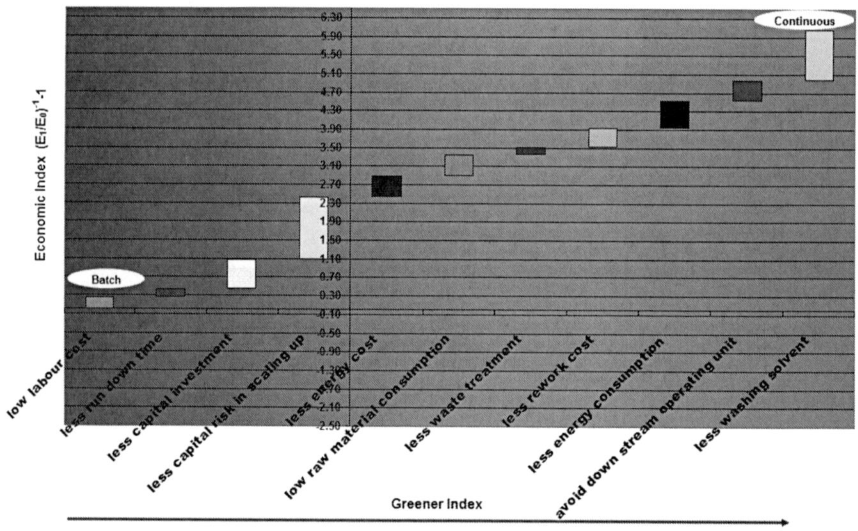

Figure 4: Economic and "green" indexes of Corning® Advanced-Flow™ reactor.

Assuming further successful development and penetration of continuous-flow technology based on employment of microchannel reactors, the potential for increased energy efficiency for chemical manufacturing processes and CO_2 emission reduction can be significant.

Internal analysis was performed in order to estimate the scale of potential environmental impact form continuous-flow microreactor-based technology penetrating pharmaceuticals and crop protection markets. The analysis utilizes results from a market forecast and technology penetration model developed internally and averaged industrial numbers on energy intensity for pharmaceuticals and crop protection segments available from various public sources. Our internal market and technology assessment revealed that it is possible that by 2015 approximately 1.7 billion tones of active ingredients (measured by mass throughput) in pharmaceuticals and crop protection segments could be produced using continuous-flow processing. The average energy intensity of pesticide active ingredients production is about 350,693 Btu (370 MJ) per kg of the final product [10]. For pharmaceuticals, energy intensity of API production on average can reach levels between 47,391 Btu/kg and 284,346 Btu/kg [11]. The calculation of the final product volumes was made with an assumption of 85:1 and 28:1 of total throughout to final product conversion ratios for pharmaceuticals and crop protection segments correspondingly. The applied conversion ratios do not consider any impact from continuous-flow process use, but represent batch process average industrial ratios in accordance with our internal estimate. Based on the National Energy Foundation, conversion rates for EU-27 grid electricity [12], from utilization of 100,000 Btu, the amount of CO_2 emitted is approximately equal to 12 kilograms. Referring to our internal energy efficiency benefit analysis and possible scenario of the future increase in use of continuous-flow production process by pharmaceuticals and crop protection segments, and assuming only electricity being used as a source of energy, by 2015 annual energy use by these two industry sectors may drop down by 118 quadrillion Btu, which is approximately equivalent to annually emitted 14.3 billion metric tones of CO_2 globally. The calculation, of course, provides a very rough estimate of the potential environmental effect of microreactor-based process intensification, but it allows getting a better idea of the scale and importance of the potential impact from the continuous flow technologies penetrating chemical industry segments, which traditionally favour batch manufacturing. Also, it allows us to mention that additional economical benefits from the employment of the continuous-flow production process can be achieved in the countries where the government imposes CO_2 taxes on the chemical producers.

5 Conclusion

Summarizing all benefits of Corning® Advanced-Flow™ reactor technology for process intensification, it is possible to conclude, that for the chemical process itself, Corning technology offers an increase of reaction rates and yields, enables new reaction paths development and simplifies downstream processes. At the same time, the technology makes a significant contribution toward chemical process safety improvement, reducing unstable intermediate acumen, and enabling in some cases “high-energy” chemistry. Taking into account all of the benefits mentioned above, it is clear that the technology has the potential to significantly increase economic efficiency of chemical production, lower operation costs due to decreased energy utilization and raw material consumption, improve safety, lower labor cost and help minimize risks associated with process scale-up. Simultaneously, the described technology can contribute to intensification of many chemical manufacturing processes while decreasing of negative environmental impact.

References

[1] Schilling C., A Balanced Diet For The Chemical Industry, 14 October 2010, www.forbes.com/2010/10/13/dupont-basf-feedstocks-technology-renewable-chemicals_2.html

[2] United Stated Department of Energy, The Industrial Technology Program (ITP), Online, www.eere.energy.gov/industry/chemicals

[3] Botschek P., Director Energy & HSE European Chemical Industry Council (Cefic), Renewable Resources – perspectives for the chemical industry, *Presentation at Renewable Resources – Sustainable Future Forum*, Helsinki region, Finland, November 2006.

[4] Bieling H.-H., Chemical reaction - an energy-intensive industry finds the solution in CHP, *Cogeneration & Onsite Power Production Magazine*, 8 (2), March 2007.

[5] Woehl P., Efficient Processing with Corning® Advanced- Flow™ Reactor Technology, *Proceedings of Industrial Green Chemistry Workshop*, Mumbai, India, December 2009.

[6] Chevalier B., Lavric E. D., Cerato-Noyerie C., Horn C. and Woehl P., Microreactors for industrial multi-phase applications. Test reactions to develop innovative glass microstructure design, *Chemistry Today*, 26, 1-4, 2008.

[7] Braune S., Poechlauer P., Reintjens R., Steinhofer S., Winter M., Olivier L., Guidat R., Woehl P., Guermeur C., Selective nitration in a microreactor for pharmaceutical production under cGMP conditions. *Chimica Oggi*, 27(1), pp. 26-29, 2009.

[8] Buisson B., Donegan S., Wray D., Parracho A., Gamble J., Caze P., Jorda J., Guermeur C., Slurry hydrogenation in a continuous flow reactor for pharmaceutical application, *Chimica Oggi*. 27(6), pp. 12–16, 2009.

[9] Pissavini S., Teaming up Chemistry and Chemical Engineering for "Greener" Processes and Improved Economics, *Presentation at 14th NIChE Conference on Micro-Reactor Technologies*, Gaithersburg, MD, USA, September 2009.

[10] Audsley E., Stacey K., Parsons D.J., Williams A.G., Estimation of the greenhouse gas emissions from agricultural pesticide manufacturing and use. Prepared for Crop Protection Association, Cranfield University, Bedford, UK, August 2009.

[11] Slater C.S., Savelski M.J., Hesketh R.P., Green Engineering Opportunities in the Pharmaceutical Industry, *Proceedings of Great Lakes Regional Pollution Prevention Roundtable Conference*, New York City, NY, USA, August 2005.

[12] National Energy Foundation, CO2 calculator, Online, www.nef.org.uk /greencompany/co2calculator.htm

Section 2
Smart processing technology for sustainability
(Special session organised by Prof. M. Naito)

Smart powder processing for energy and environment

M. Naito, H. Abe & A. Kondo
Smart Processing Research Center,
Joining and Welding Research Institute, Osaka University, Japan

Abstract

Smart powder processing stands for novel powder processing techniques that create advanced materials with minimal energy consumption and environmental impacts. This process is also important from the viewpoint of sustainability. Particle bonding technology is a typical smart powder processing technique to make advanced composites. In this paper, its application examples for clean energy, energy-saving and novel recycling will be explained. For clean energy issues, the role of microstructure control of the electrodes for higher performance of Solid Oxide Fuel Cell (SOFC) by using particle bonding will be explained. From the viewpoint of energy-saving, a new method to produce fiber reinforced fumed silica compacts by making use of particle bonding will be introduced. It will be explained that the compacts have very low thermal conductivity with high porosity at high temperature. Furthermore, by carefully controlling the bonding between different kinds of materials, effective separation of elemental components can be achieved. It leads to the development of a novel technique for recycling advanced composite materials and turns them to high-functional applications. In this paper, its application for recycling of glass fiber reinforced plastics (FRP) will be explained.
Keywords: smart powder processing, particle bonding, nanoparticles, composite particles, energy-saving, recycling, SOFC, FRP.

1 Introduction

Smart powder processing stands for novel powder processing techniques that create advanced materials with minimal energy consumption and environmental impacts. This process is also important from the viewpoint of sustainability.

WIT Transactions on Ecology and the Environment, Vol 154, © 2011 WIT Press
www.witpress.com, ISSN 1743-3541 (on-line)
doi:10.2495/CHEM110061

Particle bonding technology is a typical smart powder processing technique to make advanced composites. The technology has two main unique features. Firstly, it creates direct bonding between particles without any heat support or binders of any kind in the dry phase. The bonding is achieved through the enhanced particle surface activation induced by mechanical energy, in addition to the intrinsic high surface reactivity of nanoparticles. Using this feature, desired composite particles can be successfully fabricated. The second feature of this technology is its ability to control the nano/micro structure of the assembled composite particles. As a result, it can custom various kinds of nano/micro structures and can produce new materials with a simpler manufacturing process.

Fig. 1 shows the examples of unique microstructure created by the particle bonding technique, which led to various kinds of applications. The a) of Fig. 1 shows SEM image of multi-layered composite particle for the drug delivery systems (DDS) [1]. The b) is a TEM photograph of tungsten carbide (WC) particles bonded by nano-sized platinum (Pt) particles and carbon nanoparticles. By bonding these components to make composite catalyst, the amount of Pt used for Polymer Electrolyte Fuel Cell (PEFC) could be reduced down to about 25% compared to that using commercially available Pt/C catalyst [2].

The c) of Fig. 1 shows a SEM photograph of the anode microstructure of Solid Oxide Fuel Cell (SOFC) created by using composite particles as starting powder materials. In this paper, the role of microstructure control of the electrodes for higher performance of SOFC by using particle bonding process

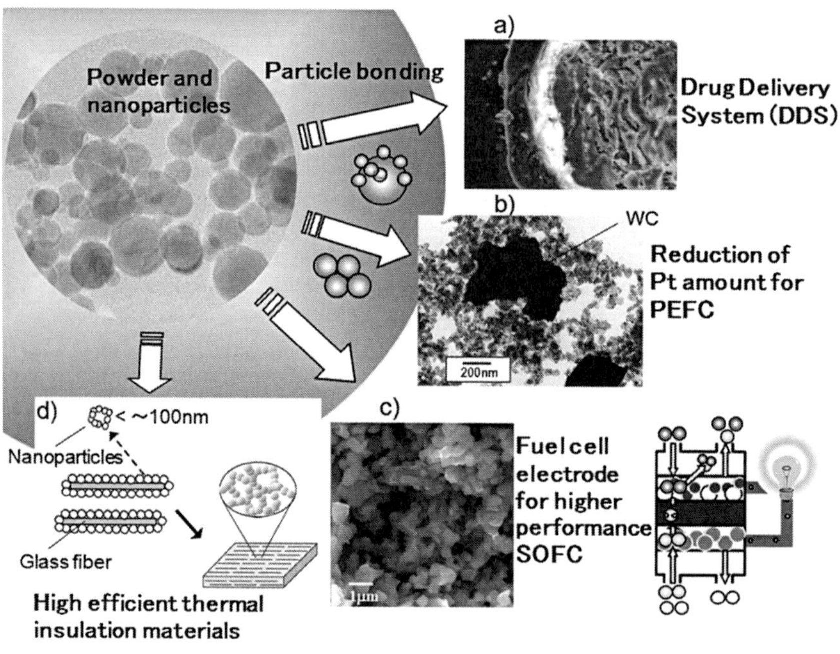

Figure 1: Unique microstructure created by particle bonding technique.

will be explained. From the viewpoint of energy-saving, interest in thermal insulation materials has been globally, because escalating energy costs signified the importance of efficient thermal insulation. The d) of Fig. 1 shows a new method to produce fiber reinforced fumed silica compacts by making use of particle bonding [3, 4]. In this paper, it will be explained that the compacts have very low thermal conductivity with high porosity at high temperature.

Furthermore, by carefully controlling the bonding between different kinds of materials, effective separation of elemental components can be achieved. It leads to the development of a novel technique for recycling advanced composite materials and turns them to high-functional applications. In this paper, its application for recycling of glass fiber reinforced plastics (FRP) will be explained [5, 6].

2 Structural control of SOFC electrodes

SOFC is a promising candidate for power generation in the 21st century because of its high energy efficiency and clean exhaust. Current R&D efforts focus on reducing its production cost and increasing the long-term stability of cells and stacks by lowering its operation temperature without losing power density. Prefabrication of the composite particles followed by electrode forming using particle bonding process is an ideal way to go, especially for controlling the microstructure of composite electrodes. Recently, we successfully fabricated various kinds of composite particles such as large core-particles coated with nanoparticles [7] and inter-dispersed composite mixture consisting of several kinds of nanoparticles using the particle bonding technique [8].

Nickel-yttria stabilized zirconia (Ni-YSZ) is the most widely used SOFC anode material due to its excellent electrochemical properties at high temperatures. The electrochemical reaction (hydrogen oxidation) takes place at the triple-phase boundary (TPB) where Ni, YSZ and fuel gas meet. The reaction rate strongly depends on the catalytic activity of anode materials and the TPB length. Since the former significantly decreases with decreasing operation temperature, the latter must be increased as much as possible in the limited effective electrode volume to keep high electrochemical performance even at lower temperatures. For the TPB enlargement, the anode microstructure such as size and arrangement of Ni and YSZ must be controlled precisely.

Fig. 2 shows the process to fabricate SOFC anode using coated composite particles, i.e. NiO large core-particle coated with YSZ nanoparticles [5]. The composite particles provided well-organized electrode microstructure as shown in the figure, and resulted in good electrochemical performance. As a result, a prototype SOFC power plant system with a capacity of about 1kW was fabricated using the anode made by the coated composite particles and has been in services at an operating temperature of 700°C [5].

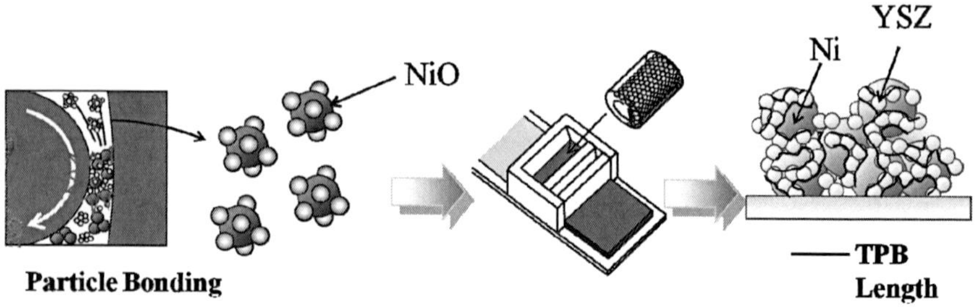

Figure 2: Unique microstructure created by particle bonding technique.

3 Application of particle bonding for thermal insulation

The second example is to make high efficient thermal insulation materials [3, 4]. Interest in thermal insulation materials has been intensified globally, because escalating energy costs signified the importance of efficient thermal insulation. In this study, nanoparticle bonding process was used to make composite fibers coated with porous fumed silica layer in the dry phase. Fig. 1 d) shows the proposed dry processing method to fabricate fumed silica compact by using composite fibers [3]. Fiber glass composites porously coated with silica nanoparticles were fabricated at the first stage and then compacted into a board by dry pressing. The composites were produced by a particle bonding process without collapsing the fiber glass and nano-scale pores made by the fumed silica. The proposed method had the advantage of preventing contacts between fibers in the compacts due to the existence of coating layer. In addition, since fumed silica was fixed on the fibers, particle segregation rarely occurred during forming. Therefore, highly uniform dispersion of fibers in the compact could easily be achieved.

Table 1: Thermal conductivity of the fibrous fumed silica compacts prepared by particle bonding process.

Specimen	Density [kg/m³]	Porosity [%]	Thermal Conductivity [W/(m·K)]	
			@100℃	@400℃
A	459	81.2	0.0266	0.0269
B	485	80.1	0.0266	0.0282

Table 1 shows the thermal conductivity of compact specimens with 80% porosity at 100°C and 400°C [3]. They were lower than molecular conductivity of still air (0.03 W/mK at 100°C, 0.05 W/mK at 400°C) and at the same level as those obtained from silica aerogel and fumed silica compacts. These results indicated that the obtained compacts had nano-scale porous structure. The remarkable attribute of composite fibers was achieving very low thermal conductivity with a relatively large amount of glass fibers. Mechanical strength of the compacts depends on their apparent density determined by the compressive strength. In this case, fracture strength ranged from 0.4 to 1.6 MPa, corresponding to apparent densities from 400 to 480 kg/m^3, could be obtained. This made it possible to machine the compacts for various applications.

The present study demonstrated that the dry powder processing method as shown in Fig. 1d) provided fibrous fumed silica compacts with mechanical reliability and efficient thermal insulation. The specific feature of this method was to apply effective mechanical processing for making fumed silica fiber composites, instead of conventional mixing techniques. The powder mixture consisting of composite fibers was in a good mixing state and resulted in fumed silica compact with well-dispersed fibers. The existence of fumed silica layer on the glass fiber could prevent direct fiber-fiber contacts in the compact which avoided solid thermal transport through the fibers even at high fiber loadings. Thus, as mentioned above, fumed silica compacts with efficient thermal insulation and good mechanical strength were achieved in this study [3].

Furthermore, the thermal conductivity of fumed silica compacts at higher temperature could also be kept at lower value by adding SiC powders as an opacifier. It was found that thermal insulation compacts made of powder mixture consisting of fumed silica: glass fiber: SiC at a mass ratio of 70:10:20 prepared by the particle bonding technique could achieve a thermal conductivity of 0.04 W/mK at 600°C [9]. By changing the kind of nanoparticle additives, it was expected that thermal conductivity of the compacts could be kept lower at even higher temperatures by particle bonding process in the future.

4 Novel recycling for lower energy and sustainability

Fig. 3 shows the concept of smart recycling of composite materials by using the particle bonding and disassembling between different kinds of materials [5]. From the conventional recycling standpoint of view, each element of waste composite material must be returned back to its original state for repeated usage. It is acceptable when the purpose of the recycling is to recover only valuable element from the waste materials. However, it needs high recycling costs and the obtained element has lower quality than the virgin material. As a result, the recycling process cannot be practically used. On the contrary, the proposed recycling concept does not aim to obtain each original element, but develops further advanced materials using disassembled blocks of the waste composite materials. In this case, how to apply bonding and disassembling the waste materials is the very key issue. As shown in Fig. 3, recycling waste back to its intermediate structure and then assembling it with another material to make further advanced materials would be more energy efficient than reclaiming the original elements. This concept will be a basis for the next generation of recycling system for advanced materials. In this paper, an approach to recycle glass fiber reinforced plastics (FRP) will be explained as follows:

FRP is a typical composite material having the advantages of lightweight, high strength and high weather resistance. Therefore, it has been used in various applications including boats, bath tubs, and building materials. Its production volume reached 460,000 tons in Japan in 1996, but decreased gradually since then. However, the volume of waste FRP has increased every year. So far, almost all of the waste FRP has been incinerated or disposed in landfill. Only 1–2% of the waste FRP is recycled as cement raw material or additives for concrete. Japan Reinforced Plastics Society started producing cement recycled from FRP in 2002. The incineration of FRP has problems of low calorific values on burning, and its residue needs to be disposed.

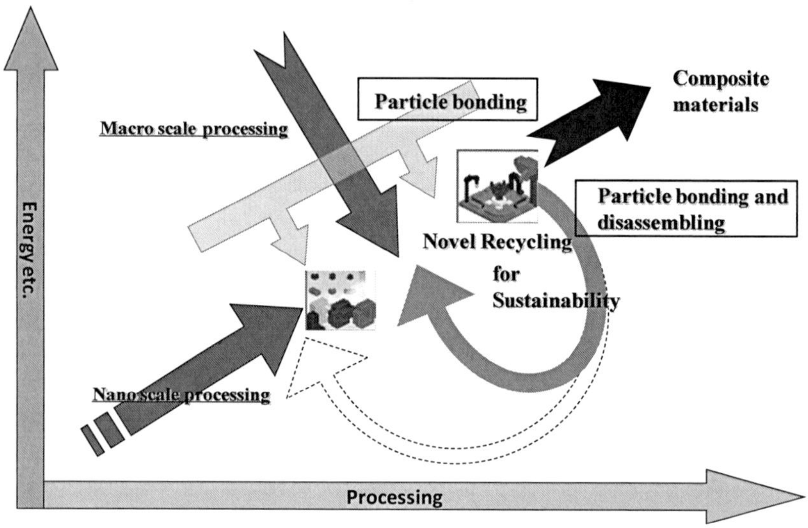

Figure 3: Concept of smart recycling for composite materials based on particle bonding principle.

FRP usually contains 40-50% of calcium carbonate filler and 20-30% of glass fibers. These materials must be recycled through simple and low energy process for profits. Therefore, we aimed to develop new recycling method to make advanced materials from the waste FRP based on the concept as shown in Fig. 3. Fig. 4 shows the concept of an innovative recycling process of FRP proposed by the authors [5]. It consisted of two unit processes. First, FRP was separated into glass fibers and matrix resins, and then, the surface of separated glass fibers was coated by low cost nanoparticles. The coated composite glass fibers would be compacted to make porous materials as shown in Fig. 1 d). High functional materials having the properties of very low thermal conductivity, light weight, and easy machining are expected to obtain by applying the new process shown in Fig. 1 d).

The waste FRP chip crushed down to about 1 cm was processed by an attrition-type mill, which applied similar mechanical principle to that of particle bonding process. When strong shear stress was applied to the chip layers for surface grinding, glass fibers began to separate from matrix resins on the chip surfaces. As a result, all glass fibers were effectively separated from other matrix components.

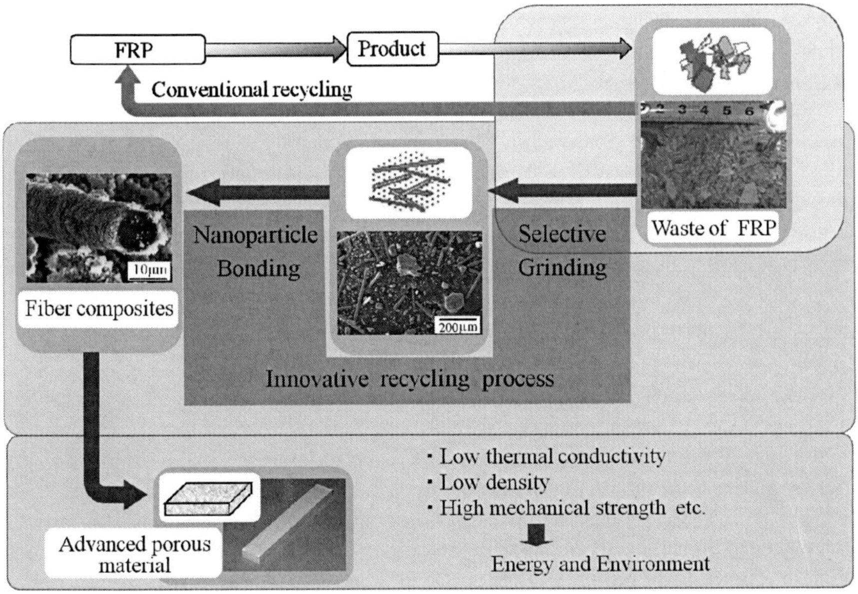

Figure 4: Smart recycling process of FRP.

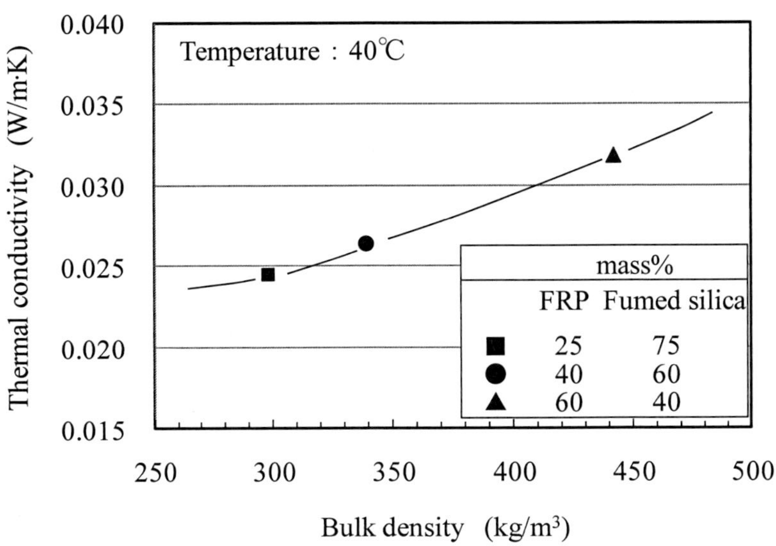

Figure 5: Relationships between thermal conductivity and bulk density of the compacts.

Then, the surface of glass fibers and that of matrix components were mechanically bonded with nanoparticles. Fig. 4 shows the glass fiber composite particles coated with fumed silica nanoparticles. It shows that the surface is covered by nanoporous layer. The board was compacted with the mixture of glass fiber composites and matrix components by dry pressing. The board has relatively high fracture strength. Therefore, it was easy for machining into various shapes.

Fig. 5 shows the relationship between thermal conductivity and bulk density of the compact [9]. When fumed silica mass percentage increases, the bulk density decreases, thus leads to extremely lower thermal conductivity of the compact as shown in Fig. 5. These results suggest that the board will be used as a new material such as building material in the future.

5 Conclusions

In this paper, particle bonding process was explained as a typical example of smart powder processing. This process enabled us to develop new composite materials. Based on its principle, the electrodes for SOFC and the high performance thermal insulator were developed. This process is also applicable for new recycling process for waste composite materials. By using this process, the waste materials will be turned into other advanced materials. It is believed that dry particle bonding technology can open the doors for various kinds of smart powder processing applications in the future.

References

[1] Fukumori, Y., Ichikawa, H., Uemura, T., Sato, K., Abe, H. & Naito, M., Process performance of dry powder coating for preparing controlled release microcapsules by a High speed mixer. *Proceedings of 8th International Symposium on Agglomeration, Bangkok, Thailand, March 16-18,* pp. 31-38, 2005.

[2] Munakata, H., Tashita, T., Kanamura, K., Kondo, A. & Naito, M., Development of low-platinum catalyst for fuel cells by mechano-chemical method. J. Soc. Powder Technol., Japan, **48(6)**, in press, 2011.

[3] Abe, H., Abe, I., Sato, K. & Naito, M., Dry powder processing of fibrous fumed Silica compacts for thermal insulation. J. Am. Ceram. Soc., **88**, pp. 1359-1361, 2005.

[4] Abe, I., Sato, K., Abe, H. & Naito, M., Formation of porous fumed silica coating on the surface of glass fibers by a dry mechanical processing technique. Advanced Powder Technology, **19**, pp. 311-320, 2008.

[5] Naito, M., Abe, H., Kondo, A., Yokoyama, Y. & Huang, C.C., Smart powder processing for advanced materials. KONA Powder and Particle Journal, **27**, pp. 130-141, 2009.

[6] Kondo, A., Abe, H., Isu, N., Miura, M., Mori, A., Ohmura, T. & Naito, M., Development of light weight materials with low thermal conductivity by making use of waste FRP. J. Soc. Powder Technol., Japan, **47**, pp. 768-772, 2010.

[7] Fukui, T., Murata, K., Ohara, S., Abe, H., Naito, M. & Nogi, K., Morphology control of Ni-YSZ cermet anode for lower temperature operation of SOFC. J. Power Sources, **125**, pp. 17-21, 2004.
[8] Sato, K., Abe, H., Misono, T., Murata, K., Fukui, T. & Naito, M., Enhanced electrochemical activity and long-term stability of Ni-YSZ anode derived from NiO-YSZ interdispersed composite particles. J. European Ceram. Soc., **29**, pp. 1119-1124, 2009.
[9] Ohmura, T., Abe, I., Ito, Y., Sato, K., Abe, H. & Naito, M., Development and evaluation of characteristics of nanoporous materials. J. Soc. Powder Technol., Japan, **46**, pp. 461-466, 2009.

Advanced technique to reduce emissions of fine particulate matter using ultrasounds

B. Bergmans[1], M. Dormann[2], F. Idczak[1], S. Petitjean[1], D. Steyls[2] & B. Vanderheyden[2]
[1]*Institut Scientifique de Service Public (ISSeP), Belgium*
[2]*Centre for Research in Metallurgy (CRM), Belgium*

Abstract

Ever stricter air quality standards force all activity sectors to take measures to reduce the emission of particles. In the past, dust measurements at stack mostly referred to Total Suspended Particles (TSP), but nowadays there are more and more concerns about the finest fractions. Indeed, when inhaled, small particles can go deeply into the lungs and are thus the most toxic ones. Particulate Matters (PMs) are generally classified in different fractions regarding their size (PM_{10}/$PM_{2.5}$/PM_1/UFP). A new European standard (ISO 23210) dealing with PM measurements in stacks has been released in 2009 and abatement impositions are expected within short delays. In the framework of various research projects the Centre for Research in Metallurgy (CRM) and its partners are studying the possibility to comply with stricter PM regulation by improving the collection efficiency of existing dedusting systems, in order to avoid expensive investments in additional abatement means. One idea is to agglomerate fine dust particles together or onto coarser ones by applying ultrasounds to the dust laden waste gas flow, upstream of the dedusting unit. This should facilitate the fine dust collection, whatever the considered dedusting technique. The feasibility of this concept has been studied at laboratory scale on a specific pilot for different industrial processes. Specific measurement techniques have also been developed to measure in real time the effects of ultrasounds on the various size fractions. For the moment the abatement reaches already up to 40% for the smallest particles ($PM_{2.5}$ and PM_1), making this system very attractive. Optimization of the method is still under progress. Present activities deal with ultrasounds frequency, acoustic chamber design and up-scaling modelling.
Keywords: ultrasounds, abatement, particles, PM, pre-treatment.

WIT Transactions on Ecology and the Environment, Vol 154, © 2011 WIT Press
www.witpress.com, ISSN 1743-3541 (on-line)
doi:10.2495/CHEM110071

1 Introduction

PM (Particulate matter) emitted by human activities are responsible for many respiratory diseases in Europe and other industrial countries. Reductions in statistical mean life expectancy that can be attributed to the contribution of fine PM are estimated at 8 months for the European population. The figure for Belgium is one of the highest ones in all European countries with 14 months and reaches up to 2 years in some highly polluted areas [1].

PM10 emissions from large industrial plants should already be reported to fulfil with the European IPPC directive [2]. Impositions on smaller fractions (PM2.5, PM1, and UFP) are also under discussion. A new European standard dealing with PM measurements in stack has been released in 2009 [3]. Measurement campaigns will thus be imposed by environmental agencies and abatement impositions are expected within short delays.

The purpose of this study is to demonstrate the possibility to comply with stricter PM regulations by improving the collection efficiency of existing dedusting systems using ultrasounds technique, in order to avoid expensive investments in additional abatement means.

2 Material and method

2.1 Principle of the concept

The idea is to agglomerate fine dust particles together or onto coarser ones by applying ultrasounds to the dust laden waste gas flow, upstream of the dedusting unit. This should facilitate the fine dust collection, whatever the considered abatement techniques. Due to their lower inertia, the finest particles are selectively put into vibration by the ultrasonic pressure waves induced by the ultrasounds, resulting in an increase in the number of collisions approving the phenomenon of agglomeration (see figure 1).

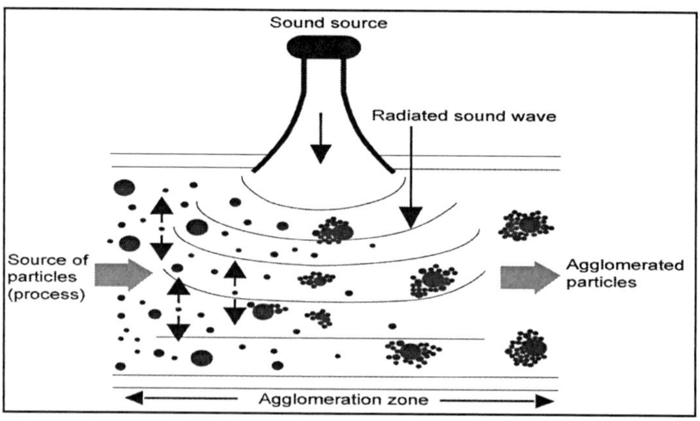

Figure 1: Principle of in-duct acoustic agglomeration of fine dust.

According to a former study carried out on coal combustion fumes by CSIC [4], PM abatement improvements of 40% (in mass) can be expected.

2.2 Description of the pilot installation

A specific pilot installation (see figure 2) has been built to allow tests on artificial fumes with predetermined composition and concentration of dust. Specific attention has been put in order to obtain regular and homogeneous dust dispersion. Gas composition, flow rate, temperature and moisture can also be controlled to be as close as possible to real plant conditions.

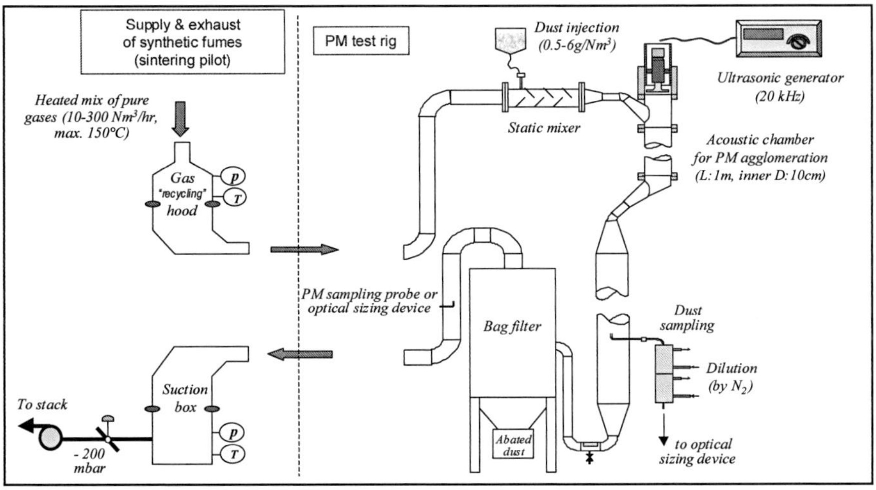

Figure 2: Diagram of the fine dust agglomeration test rig.

The agglomeration chamber is a 1 metre long tube (inner diameter 10 cm) with a 20 kHz ultrasounds (US) generator and transducer (sonotrode). The maximum acoustic power achieved is 165 dB. The only material suitable for the ultrasounds emitter in this range of frequencies and powers is titanium, due to its low weight on resistance ratio and its rigidity. A bag filter is used to check the induced dust abatement improvement.

2.3 Description of the measurement method

Real time monitoring of particles is necessary to demonstrate the effect of ultrasounds on dust agglomeration. This measurement has been performed using a quantitative dilution system (ejector principle) linked to an optical sizing device (measuring range: from 0.3 to 20 μm). A field oriented prototype has been designed, taking usual constraints of emission measurements and secondary particles formation optimization into account. The validation of the method has been performed in a previous study [5].

In order to assess the effect of dust agglomeration on bag filter yield, upstream and downstream measurements were performed simultaneously. Dilution with a factor of 100 is required upstream to avoid the coincidence error (in the optical sizer) due to the high dust concentration. After the bag filter, the use of a diluter is not necessary because the dust concentration is much lower.

Dust collections on Teflon® filters (for weighing analysis and microscope observation) have also been performed to confirm the result of the optical measurements.

3 Results and discussion

3.1 Effect of ultrasounds acoustic power on agglomeration yield

Figure 3 shows the effect of acoustic power on dust agglomeration. The range of acoustic power tested varies from 157 to 162 dB. Ultrasounds have a significant effect around 160 dB which confirms the value expected from literature [6]. Results also confirm that the effect of acoustic power is not proportional, there is a value below which no significant results are obtained, followed by a zone with high efficiency and then only very small gain are obtained when increasing the acoustic power. As acoustic level is directly related to the electrical power consumption this parameter is important from an economic point of view.

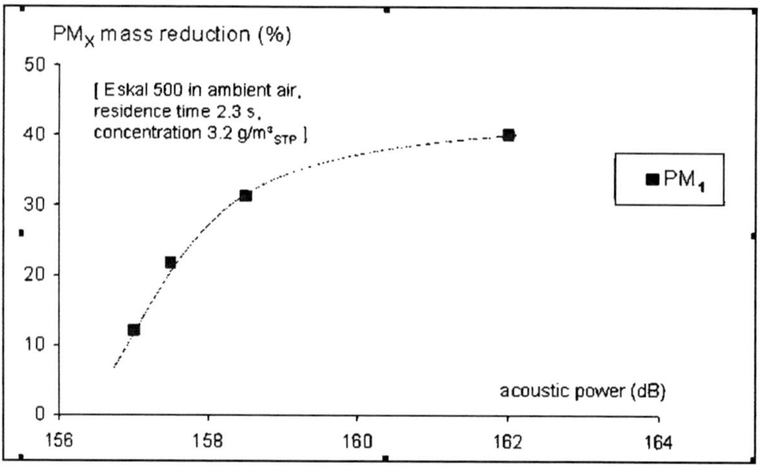

Figure 3: Effect of ultrasounds power source on agglomeration.

3.2 Effect of residential time in ultrasounds chamber on agglomeration yield

By increasing the ultrasounds treatment time from 0.6 to 2.3 seconds, a sharp increase of the PM agglomeration efficiency is observed (see figure 4). No significant improvement effect is expected with residence time above 3 seconds.

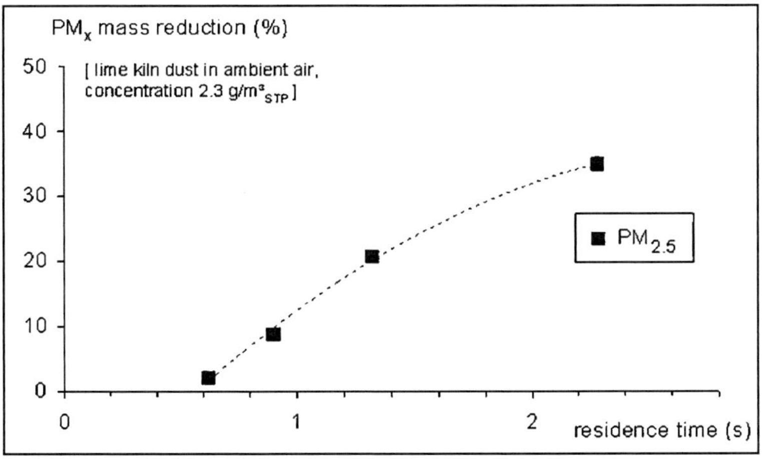

Figure 4: Effect of the residential time on agglomeration.

3.3 Effect of dust concentration on agglomeration yield

The effect of dust concentration is presented in Figure 5. A higher concentration in the range of 1.5 to 5 g/m³ of dust helps to some extent to the particles agglomeration. This phenomenon is simply related to a higher probability of collision at higher concentration.

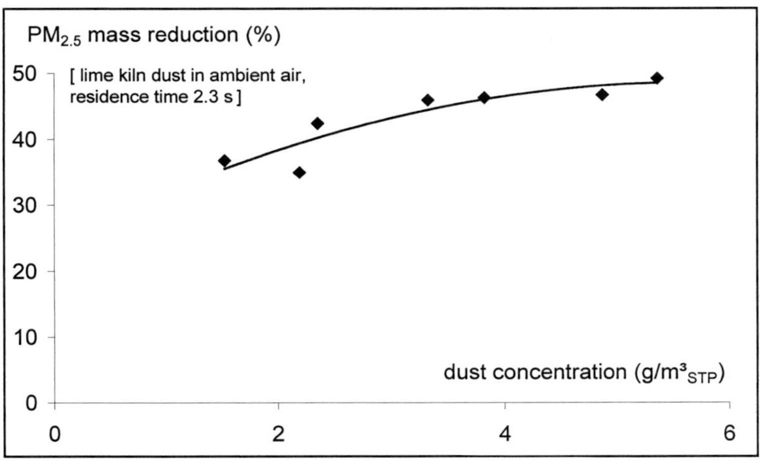

Figure 5: Effect of dust concentration on agglomeration.

3.4 Effect of ultrasounds on specific class size

It is interesting to notice that the mass reduction of PM1 is somewhat higher than that of PM2.5 or coarser ones (see Table 1), whatever the acoustic power or

Table 1: Effect of ultrasounds on dust from different origins.

	Eskal 500	ZnO	Lime kiln	Sinter plant
Origin	Commercial	Commercial	Sampling	Sampling
v/v% < 1 µm	4.5	0.7	6.0	2.1
v/v% < 2.5 µm	20.2	4.7	11.2	14.3
v/v% < 10 µm	97.8	22.6	53.0	17.4
Abatement PM1	30%	42%	26%	25%
Abatement PM2.5	21%	32%	20%	21%

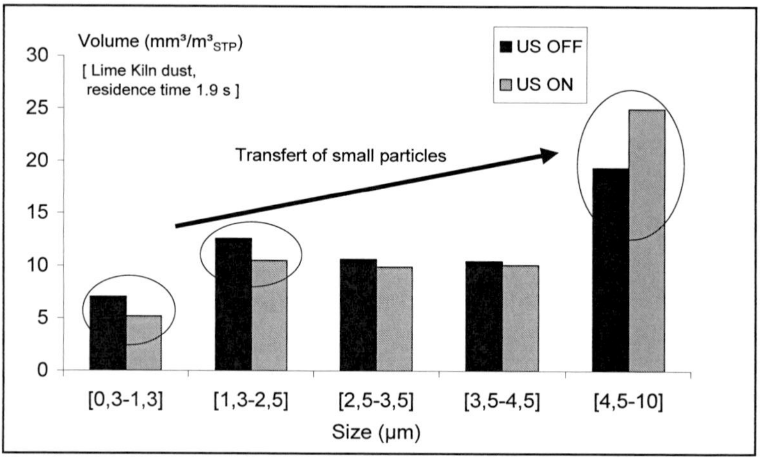

Figure 6: Effect of ultrasound upstream of bag filter.

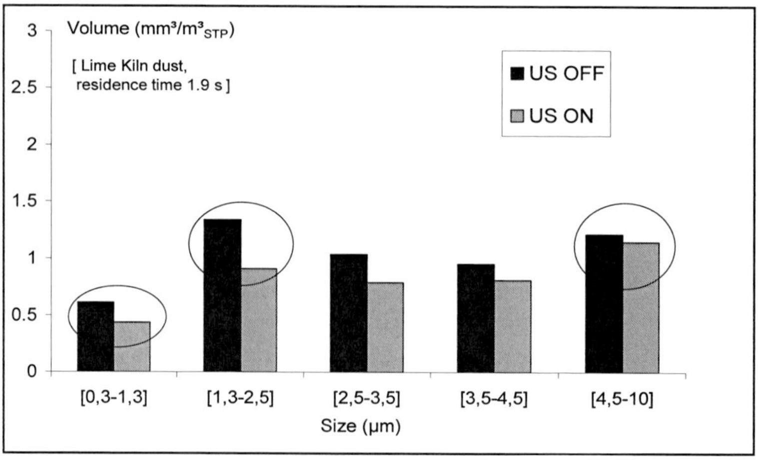

Figure 7: Effect of ultrasounds downstream of bag filter.

residential time. This effect is due to the high frequency (20 kHz) applied by the sonotrode which selectively puts the finest particles into vibration [7].

Figure 6 shows the variation induced by ultrasounds for five size ranges upstream of the bag filter. One can observe a mass transfer from small particles (< 2.5 μm) to coarser ones (> 4.5 μm) when applying ultrasounds.

Figure 7 shows the corresponding results downstream of the bag filter.

The reduction in the smaller size classes is maintained (as a direct result of the preliminary agglomeration of small particles), whereas the increase in the higher size classes is no more observed (as a result of the capture of the clustered particles by the bag filter).

3.5 Influence of dust type on agglomeration results

Samples of dust have been collected from various industrial processes or sources to verify the feasibility of this original concept to different industrial processes potentially concerned by the PM issue.

Main dust characteristics and results obtained on them are presented in table 1.

The first dust (Eskal 500) is a commercial limestone powder used as standard for the calibration of optical sizing systems and made up of hydrophobic, dense (non porous) and nearly spherical limestone particles with a very narrow size distribution (90% mass between 1 and 10 μm). Results are thus expected to be much lower with this technical powder. The three other dusts are industrial flying dusts, collected in existing dedusting systems of a plant. They have wider size distributions with particles up to 100 microns in some cases and 5 to 15% in mass below 2.5 microns. One has been sampled in steel industry sector (Sinter plant), another one in a zinc oxide production plant (ZnO), and the last one comes from a plant producing cement (Lime kiln).

From these results we can conclude that the PM agglomeration by ultrasounds is a rather universal technique as it is able to significantly reduce the number of small particles whatever the type of dust present in the fume. Contrary to what has been expected, results remain also correct for the technical Eskal 500 powder.

3.6 Confirmation of the results using the normative measuring technique

In order to quantify the total suspended dust abatement downstream of the bag house filter, several weighing analysis of samplings collected on filters have been carried out. All sampling are performed in accordance with the requirements of the EN 13284-1 by an accredited laboratory.

Efficiency of a bag house filter increases in time, because the collected dust has also a filtering effect. To avoid an incorrect interpretation of the results, different sequences with and without ultrasounds have been sampled.

Results obtained are presented in table 2 and show a yield in mass abatement around 50% after the bag house filter when applying ultrasounds to the dust laden waste gas.

Table 2: Effect of ultrasounds on bag filter yield.

Sequence	US	Dust concentration (mg/Nm^3)
1	OFF	21.8
2	ON	9
3	OFF	23.2
4	ON	10.1
5	OFF	18.5

These tests have been performed with sinter plant dust. All parameters have been optimized (Residential time = 5s, maximum ultrasounds power, concentration of dust = 15g/Nm^3). For economical or practical reasons, these parameters can not be implemented in industrial conditions, and abatement around 20–25% is thus expected in real conditions (see Table 1). Optimization of the technique should still be done before any real field application.

3.7 Confirmation of the results by microscopy

Analyses using a scanning electron microscope (SEM) were performed on dust sampled on Teflon® filters to allow a visualization of the agglomeration process.

Figure 6 shows very well the agglomeration of small particles onto coarse ones. The first picture shows a particle collected when no ultrasounds are applied on the fume and the second one the same dust after an ultrasounds treatment.

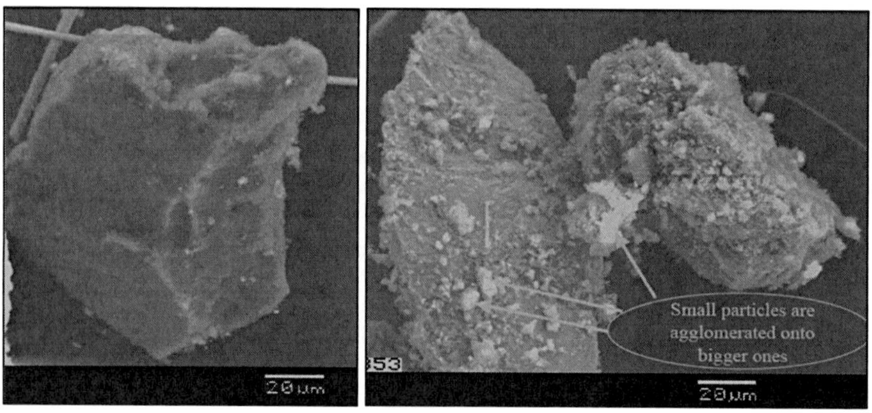

Figure 8: Visualization of the agglomeration due to ultrasounds.

4 Conclusions

The improvement of PM abatement on existing dedusting systems can be achieved by an adequate preconditioning of the fumes using acoustic technique. The obtained results show an abatement of up to 50% and the technique can be applied to a wide range of dust types and operating conditions.

Source generator and protocol have been optimized on a specific pilot installation and the different parameters (acoustic power, building material, residential time ...) are under control.

Ultrasounds are mainly acting on the smallest particles, which makes this technique very innovating. Other abatement techniques (filters, electrostatic precipitator) remove mainly the coarse fraction and are more or less inefficient on the smallest fractions.

Higher abatement yield could thus be expected on toxic species like heavy metals and Persistent Organic Pollutants (POPs), mostly concentrated on the finest particles. There could be a potentially increased interest for this technique with respect to more health oriented present and future environmental regulations. Abatement of ultra fine particles (UFP) will also be quantified in a near future.

Another advantage is that no substance is injected into the gas contrary to most preconditioning techniques based on additives, so avoiding the related negative side effects (emission or disposal issues).

Future trials aim also to check the influence of gas moisture on PM agglomeration. Indeed the presence of moisture could induce condensation phenomena on the particles modifying favourably their surface properties in terms of agglomeration efficiency and higher agglomeration in the submicron range is expected according to literature [8].

The scaling-up of this technology at industrial level requires the optimization of the acoustic chamber design in order to minimize the number of acoustic generators (for economical reasons) and the residence time (for practical reasons). Within the framework of a second research program, a new acoustic test rig able to work at higher flow rate (200 Nm^3/h) and a dedicated simulation tool are for the moment in preparation [9].

Acknowledgements

This research work is carried out with the financial support of Euregio Meuse-Rhin (Interreg III) and of the Walloon Region.

References

[1] Baseline scenarios for the clean air for Europe (CAFE) programme - Final report October 2004

[2] European directive 96/61/CE of 24/09/1996 (JOCE of 10 October 1996) concerning Integrated Pollution Prevention and Control (IPPC) and its implementation decision N(2000) 20004 - 17/07/2000 (JOCE 28 July 2000).

[3] ISO 23210:2009 Stationary source emissions - Determination of PM10/PM2.5 mass concentration in flue gas - Measurement at low concentrations by use of impactors

[4] Gallego et al., Acoustic preconditioning of coal combustion fumes for enhancement of electrostatic precipitator performance, Coal Science 1995, p. 1899-1906

[5] B. Bergmans, F. Idczak, Ph. Maetz and S. Petitjean., Real time monitoring of primary and secondary particulate matter emissions from industrial sources, 8th International Conference CEM 2007 – III – 45-50 – 2007

[6] Shaw, Acoustic particle agglomeration due to hydrodynamic interactions between monodisperse aerosols, Journ. A. Sc. 1979, p. 317-328

[7] Sergey V. Komarov et al., Numerical simulation of acoustic agglomeration of dust particles in high temperature exhaust gas, second international conference on mathematical modeling and computer simulation of metal technologies, 2002

[8] E. Riera-Franco de Sarabia et al., Investigation of the influence of humidity on the ultrasonic agglomeration of submicron particles in diesel exhausts, Ultrasonics 2003, p. 277-281

[9] Paragglo: Reduction of industrial dust pollution by acoustic agglomeration helped by chemical agglomeration, Research project RW DGO6, 2009-2011

Measurement and analysis of fine particulate matters (PM10/PM2.5) and condensable nanoparticles emission from stationary sources

H. Kamiya[1], K. Hada[1], T. Sekizawa[1], M. Yamada[1], M. Tsukada[1], W. Lenggoro[1], M. Wada[2], N. Kogure[3], Y. Yuping[4] & W. W. Szymanski[5]
[1]*Intitute of Engineering, Tokyo University of Agriculture and Technology, Japan*
[2]*Ishikawa National College of Technology, Japan*
[3]*Advanced Industrial Science and Technology, AIST, Japan*
[4]*Zhejiang Feida Environmental Science & Technology Co., Ltd., China*
[5]*Faculty of Physics, University of Vienna, Austria*

Abstract

For standardisation of sampling and measuring methods regarding in-stack measurement of PM2.5 and PM10 mass concentration in flue gas of stationary emission sources, a two stage VIS impactor (virtual impaction surface impactor) was developed in which the solid impaction surface of a conventional impactor is replaced by a space of a relatively slow moving air within a cavity of a receiving nozzle. Separation performance of the VIS impactor was examined using a model aerosol-gas stream system with a variety of dust concentrations and then exploited for the flue gas measurement in a coal combustion plant in China. Because a further stack emission is the condensable suspended particle matter (condensable SPM), two types of diluters for this measurement were designed and constructed based on England et al. and W. Lee et al.'s apparatus. Model exhaust gas with heavy metals was prepared in a laboratory scaled experimental arrangement and was mixed with clean air in both dilutor types. The effects of diluter structure, dilution ratio and residence time on the size distribution of condensable SPM are discussed.
Keywords: virtual impaction surface impactor, separation efficiency, dilutor, condensable suspended particulate matter.

WIT Transactions on Ecology and the Environment, Vol 154, © 2011 WIT Press
www.witpress.com, ISSN 1743-3541 (on-line)
doi:10.2495/CHEM110081

1 Introduction

PM2.5 (particles which pass through a size-selective nozzle with a 50% efficiency cut-off at 2,5 μm aerodynamic diameter) invades the endocyst of the lung and is considered to be one of the factors of coronary diseases such as asthma or lung cancer. High correlation of dust concentration and human health was reported [1]. ISO7708 [2] describes the ratios of particles invading the human body by breathing as a function of aerodynamic particle diameter; particles with passage through pharynx (Thoracic convention) (probability of 50%) are 10μm and particles reaching alveolus (Respirable convention) are 2.5 μm or smaller and affect particularly the high risk group (sick and infirm, or children). Recently (September 2009), the implementation of Japan environmental standards regarding PM2.5 mass concentration necessitates a standardized measurement of PM2.5 in flue gas from stationary sources. Evidently, the status of PM10 (particles which pass through a size-selective nozzle with a 50% efficiency cut-off at 10 μm aerodynamic diameter)/PM2.5 emission is necessary to investigate. Current status for PM2.5 emission from stationary sources is less recognized comparing that from mobile sources.

In ISO/TC146 "Air quality"/SC1 "Stationary source", working group, WG20, discusses the standard of PM10/PM2.5 mass concentration measurement in stack of stationary sources. ISO 23210 [3], conventional cascade real impactor method, has already been published in 2009. Conventional impactors have collection plates on which particles within a certain size interval are collected following a collision of those particles with the impaction surface. However, particle bounce [4] and re-entrainment [5] occurring particularly at high dust concentration conditions result in substantial errors regarding mass concentrations. When a glass fiber filter was used on a collection plate, John et al. [6] reported that the separation efficiency of a real cascade impactor for coarse particles larger than 2.5μm was about 70%. Based on this result, a permissible deviation of a real impactor from recommended separation efficiency in ISO 7708 was 30% in published ISO23210. However, the low separation efficiency of coarse particles results in an undesirable overestimation of PM10/PM2.5 mass concentration.

An impactor, in which the solid impaction surface is replaced by a space of a relatively slow moving air within a cavity of a receiving nozzle, was described earlier by Conner [7] and named a virtual impactor. This type of an impactor avoids in general particle bounce and re-entrainment. Since then various modifications and improvement of this type of an impactor were developed as fine particle sampler for environment. A two-stage and multiple-nozzle Virtual Impaction Surface (VIS) impactor [8] has been developed with the aim to serve as a new standard method.

A further category of the emission from stationary sources is the category of condensable SPM. Since dust collectors are generally operated in the range of 100–300°C to avoid vapor condensation, the candidates for condensable SPM in exhaust gas were in vapor state when they passed through the dust collectors. After the exhaust gas cooling in the flue gas ducting located downstream of the dust collectors mixes with atmospheric air at the exit of the stacks, thereby the condensable components in exhaust gas become particulates. In order to determine the emission behavior of condensable SPM from stationary sources, there is a need to develop a standard method of dilution and its operation conditions, such as the structure of dilutors, dilution rate and residence time in a dilutor. Various types of dilutors were designed and prepared. Number based size distributions of condensable SPM from different stationary sources were measured using each dilutor.

The objective of this study is to establish measurement methods of PM2.5/PM10 and condensable SPM from stationary sources and to accumulate such emission data. Separation efficiency and mass concentration of PM10/PM2.5 in model dust dispersed in an air flow channel were investigated using both types of impactors: the VIS and the conventional cascade impactor.

Furthermore, two types of diluters were designed and constructed based on England et al. [9] and W. Lee et al.'s apparatus [10]. Model exhaust gas with heavy metals was prepared in a laboratory scaled experimental arrangement and was mixed with clean air in both dilutor types. The effects of diluter structure, dilution ratio and residence time on the size distribution of condensable SPM were investigated and by using model exhaust gas and flue gas sample from pulverized coal combustion test facility.

2 Experimental procedure

2.1 PM10/PM2.5 mass concentration measurement by using VIS impactor and conventional (real) cascade impactor

Two kinds of separation and collection method for the measurement of PM10/PM2.5 mass concentration in stack were used in this study. Each method is in a process of becoming a standard method worked out in ISO TC146/SC1/WG20.

2.1.1 VIS impactor

The principle of operation and the key design parameters of VIS impactor are shown in fig. 1. The particle-laden gas enters the particle acceleration nozzles and accelerates depending on D_0 and the total flow rate Q_0. Only a part of the stream Q_1 leaving the acceleration nozzles enters the particle collection nozzles. Flow rate through particle collection nozzles, which is called the minor flow rate Q_1, is about 10% of the total flow rate Q_0. The remaining, larger part of the flow, the major flow Q_2, is redirected and by-passes the particle collection nozzles.

Coarse particles over a certain aerodynamic size (cut-off size) entrained to the minor flow are received by the particle collection nozzles and after passing through those nozzles collected on a filter. Fine particles smaller than this cut-off size stay in the major stream, and are directed into the next separation stage. There are two separation stages (first stage: 10 μm cut-off, second stage: 2.5 μm cut-off) in the VIS impactor presented in this study. Each stage has 6 acceleration and 6 collection nozzles. Particles with aerodynamic diameters over

10 μm are sampled on the filter of the first stage, those in the range of 10 to 2.5 μm are sampled on the filter of the second stage and those which belong to the PM2.5 size fraction are sampled on final PM2.5 collection filter.

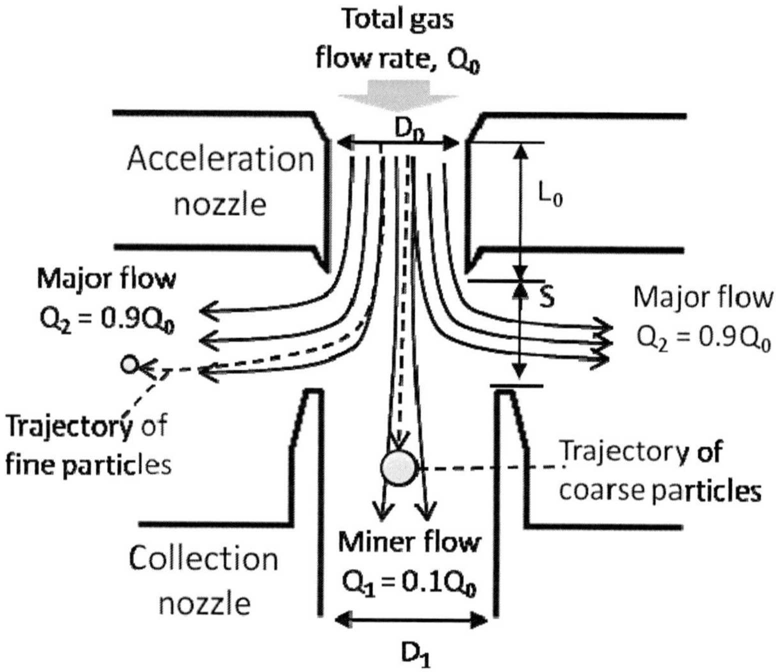

Figure 1: Principle of virtual impaction surface (VIS) impactor.

2.1.2 Conventional cascade impactor

A commercially available cascade impactor (GMU-Cascade Impactor Johnas II, Paul Goethe, Bochum, Germany) was tested in comparison with VIS impactors. The GMU impactor has two separation stages (first stage: 10 μm cut-off, second stage: 2.5 μm cut-off). The first and second stages have 6 and 12 separation nozzles, respectively. In general, coating of impaction plates with grease should limit the particle bounce and re-entrainment at the impaction plates improving the size separating performance of the impactor. However, in stacks at high temperatures in reactive atmospheres, substantial errors in PM10 and PM2.5 mass concentration measurements caused by adhesion degradation (grease) and/or weight change of coated grease are to be worried about. In this study, each impaction stage of GMU impactor with either quartz glass fiber filters or greased metal filters were used. The grease used in these experiments was fluorine corollary grease (Nichimoly, OCE@NFGS Spray).

2.1.3 PM2.5 mass concentration measurement using model aerosol generator and stack chamber

The schematic drawing of model aerosol gas stream duct simulating some industrial conditions is shown in fig. 2. JIS Z8901 Type 1 testing powder #10 (Fly ash, mass based median aerodynamic particle diameter of 7.9 μm, geometry standard deviation of 1.2 μm, particle density of 2000-2300 kg/m^3) [11] was used as model dust, and was supplied into model duct chamber by an electric dust feeder. The aggregates of dust were distributed with a mechanical brush. Mass concentration in gas stream was controlled by the feed rate of dust.

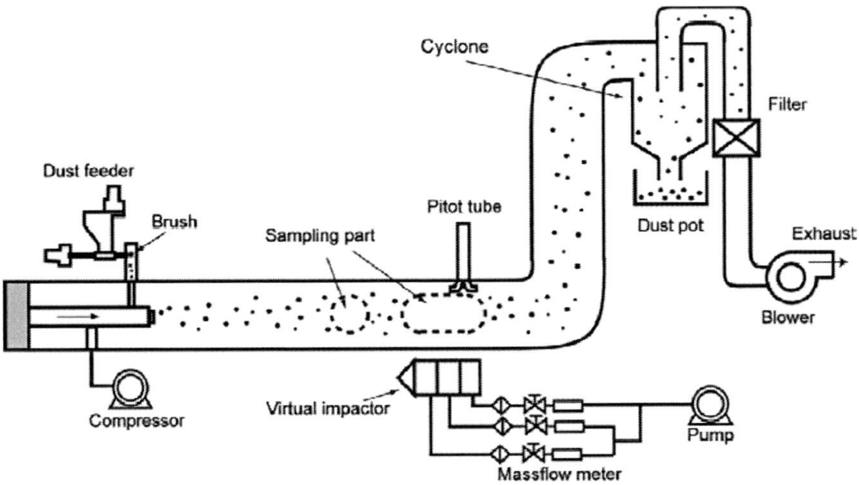

Figure 2: Schematic drawing of model testing chamber.

Since there is possibility that coarse particles in testing powder were settled out by gravity, the size distribution of dispersed model powder was measured by the following method. Dispersed model powders in gas stream were collected sampling circular filter with 37 mm in diameter at sampling part in dust chamber. The part of collected dust layer on filter was dispersed into pure water under the ultrasonic irradiation. The size distribution of collected dust was measured by a laser diffraction particle size analyzer (Shimadzu Co. Ltd., Japan, SALD-2200). In order to reduce the effect of remained particles in filter after ultrasonic irradiation on size distribution, the amount of dust layer on filter was about 50 mg, and dust layer was dispersed into water. Since coarse particles settled down during particle dispersion and transport processes, the amount of fine powder with smaller than 2.5 μm in diameter increased from about 7.0 wt% in original dust to 14 wt% in collected dust on filter.

After total dust concentration measurement, VIS impactor and/or cascade impactor was installed into model dust chamber, and the effect of total dust concentration and impactor structure on PM2.5/PM10 mass concentration determination was investigated.

2.1.4 PM2.5/PM10 mass concentration measurement from coal combustion flue gas in China

PM10/PM2.5 mass concentration was measured at the test plant of dust collectors in China. The plant combusts coal and a part of flue gas is usually introduced to an electric precipitator (ESP) and bag filters. Sampling point was the center of duct at 8.0 m downstream of bag filter unit outlet. The gas conditions were: temperature 99°C, CO_2: 8.8%, O_2: 10.9%, water content 60 g/m^3, dust concentration 120 mg/m^3, and gas velocity 13 m/s. For our measurements, ESP was stopped and a part of filters in a bag filter unit was removed in order to obtain high dust concentration in the gas. Consequently, PM10/PM2.5 was measured by virtual impactors and cascade real impactor (GMU-impactor) in different, relatively high total dust concentrations.

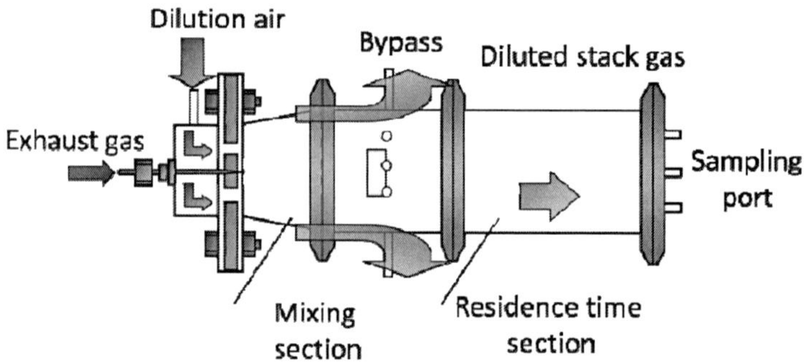

(a) ASTM type dilutor

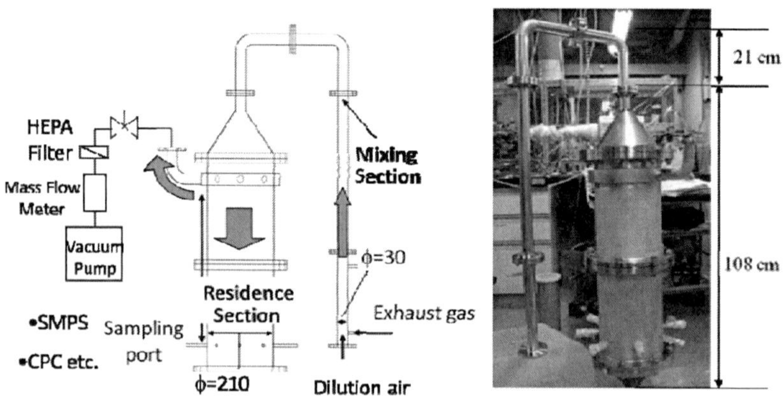

(b) CANMET type dilutor (design drawing and photograph)

Figure 3: Detail structure of two types of dilutors.

2.2 Condensable SPM measurement by using dilutors with different designed structure

2.2.1 Dilutor

ASTM [9] based type and CANMET [10] based type dilutors, shown in fig. 3, were designed. Both allow the control of the residence time by adjusting bypass gas flow rate sucked from the jacket near entrance of residence section. Detail of gas suction part was determined based on the flow patters simulated by fluid mechanic software (COMSOL) to ensure homogeneous suction for uniform flow pattern in dilutor.

2.2.2 Model flue gas with Cd vapor

An apparatus for laboratory model flue gas generation and measurement system is shown in fig. 4. $CdCl_2$, which is chosen as an example of source of emission element in waste incineration processes, was sublimed at 450°C and carried by dried air, and introduced to a dilutor with dry air dilution.

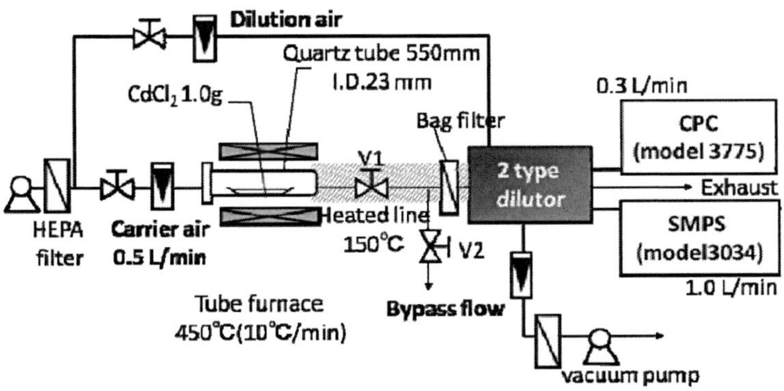

Figure 4: Experimental apparatus for the measurement of condensable SPM using model flue gas with Cd vapor.

3 Results and discussion

3.1 PM10/PM2.5 mass concentration

3.1.1 Model aerosol

Relationship between the result of measured PM2.5 concentration and the overall dust concentration is shown in fig. 5. The overall dust concentration is calculated by totaling of dust collection at each stage. In case of VIS impactor, PM2.5 concentrations are in a very good agreement with the PM2.5-line that indicates PM2.5 content over the entire dust concentration range.

In case of GMU cascade impactor, even at concentrations lower than 5 mg/m^3, PM2.5-line. Better agreement (GMU) was observed in cases when impaction plates were coated with the grease. Though the GMU cascade impactor performance was improved with the grease coating, the values were still higher than those obtained by the VIS impactor.

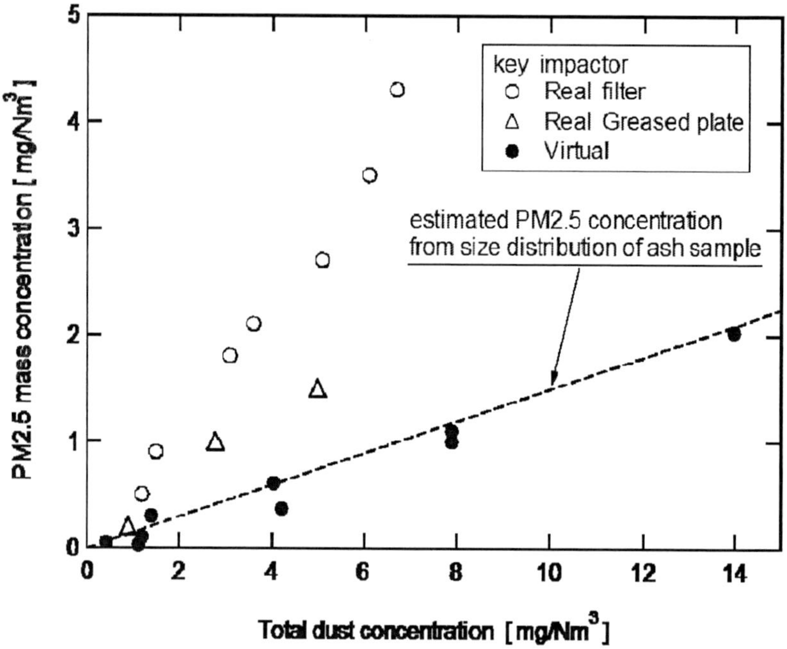

Figure 5: Relationship between total dust concentration and PM2.5 mass concentration measured by using different sampling method.

3.1.2 Coal combustion flue gas

The mass concentration of PM10 and PM2.5 was measured by two kinds of impactor in flue gas of coal combustion in China and shown in fig. 6. Since it was difficult to keep stable operation of coal dust collection, it is not enough to discuss the separation performance of both methods. However, it seems that PM2.5 mass concentration measured by GMU cascade real impactor was higher than that by VIS impactor at about almost same total dust concentration ranging less than 100 mg/Nm^3.

In order to discuss this difference, SEM images of collected dust at each stage are shown in fig. 7. Particles larger than cut-off diameter were observed which is likely caused by bounce and re-entrainment in the GMU-impactor case. In the virtual impactor case, large particles were also observed, however, the amount of coarse particles larger than 2.5 μm was very limited as can also be seen in fig. 7(b).

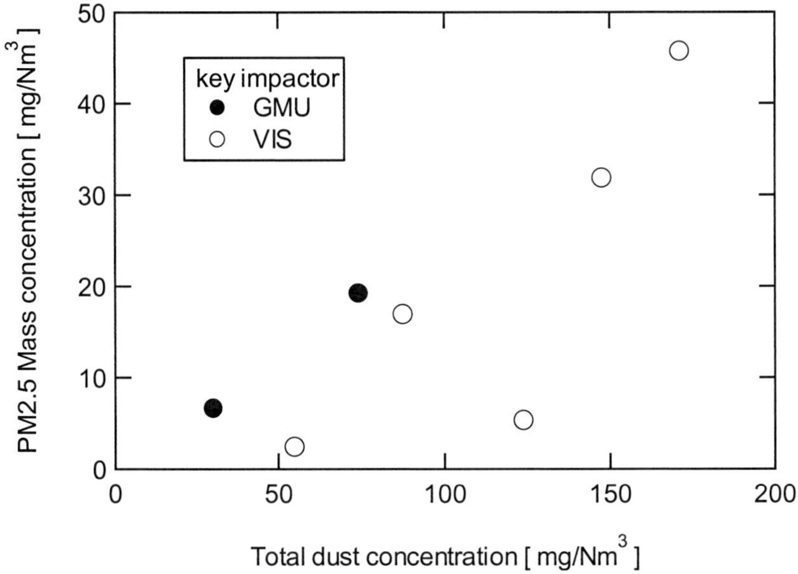

Figure 6: Relationship between total dust concentration and PM2.5 mass concentration in flue gas of coal combustion in China.

(a) GMU real impactor

(b) VIS impactor

Figure 7: SEM observation of final PM2.5 stage filter. Effect of separation method of PM2.5 on particle size of PM2.5.

3.2 Condensable SPM

The effect of dilution condition on number based size distribution of condensable SPM is shown in fig. 8. Dilution ratio and residence time were changed ranging from 10 to 30 times and from 11 to 52 s, respectively. When dilution ratio was higher than 20 times and residence time was longer than 10 s, almost same size distribution was observed.

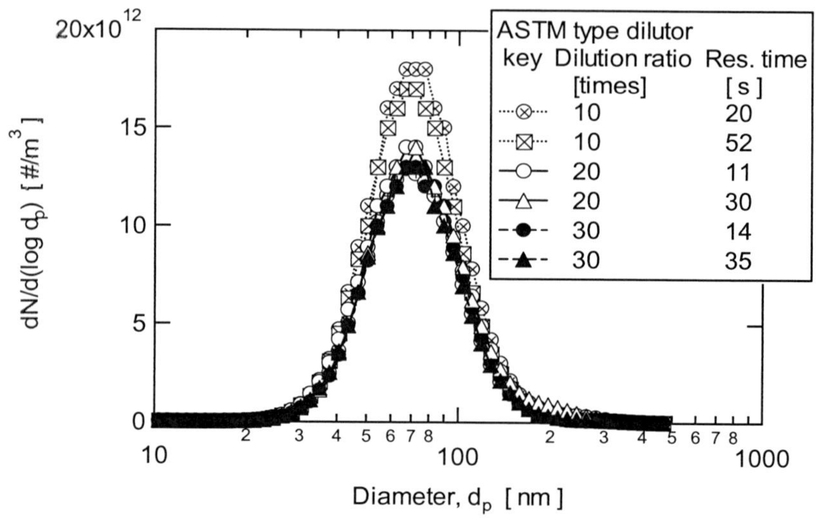

Figure 8: Effect of dilution ratio and residence time on size distribution of condensable SPM. (ASTM type dilutor).

The effect of dilutor structure on number based size distribution of condensable SPM is shown in fig. 9. Almost the same particle size distribution was obtained by using both dilutors. If dilution ratio (DR) was larger than 20 times and residence time (RT) longer than 10 s, fairly well agreements of size distribution are also obtained in two different dilutors.

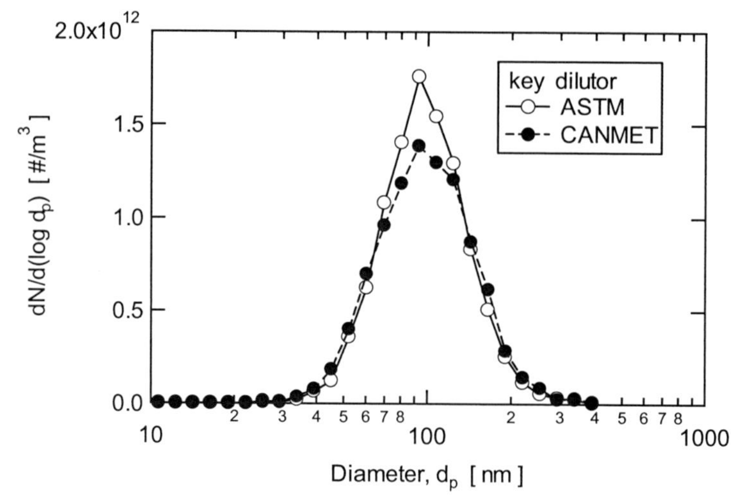

Figure 9: Effect of dilutor structure on size distribution of condensable SPM from model flue gas with Cd vapor (Dilution ratio : 40 times, Residence time : 55 s).

4 Conclusion

The conventional cascade impactor overestimated PM2.5 concentrations due to the particle bounce and re-entrainment even when adhesive coating material such as grease was applied to its impaction plates. VIS impactor showed remarkably good performance within the investigated range of PM2.5 mass concentrations and ambient parameters in the model aerosol gas flow and in the coal combustion flue gas.

With dilution ratio and residence time larger than 20 times and longer than 10 seconds, almost the same number based size distribution of condensable SPM from model flue gas with Cd vapor using two different types of dilutor was measured.

Acknowledgements

This work was supported by NEDO Intellectual Infrastructure Program, Research and Development to Promote the Creation and Utilization of an Intellectual Infrastructure (FY2005-2007), NEDO International Standardization Program (FY2008-2010), Grant-in-Aid of JSPS for Young Scientists (B) (20710028) and Grant-in-Aid of MEXT for Scientific Research on Innovative Areas (20120004).

References

[1] Dockery, D. W., C. A. Pope, III, X. Xu, et al., *New England Journal of Medicine*, 329, 1753-1759, 1993.

[2] ISO 7708: Air quality — particle size fraction definitions for health-related sampling, 1995.

[3] ISO 23210: Stationary source emissions – Determination of PM10/PM2.5 mass concentration in flue gas – Measurement at low concentrations by use of impactors, 2009.

[4] T. G. Dzubay, L. E. Hines and R. K. Stevens, *Atmospheric Environment*, 10, 229-234, 1976.

[5] K. Iinoya, S. Yuu, K. Makino and K. Nakano, *Kagaku Kogaku*, 33, 689-698, 1969.

[6] A.C. John, T.A.J. Kuhlbusch, H. Fissan, G. Bröker, K.-J. Geueke, *Aerosol Science and Technology* 37 694-702, 2003.

[7] W. D. Conner, *Journal of the Air Pollution Control Association,* 16, 35-38, 1966.

[8] T. Prasserttachato, A. Podgorski, J.H. Luckner, M. Furuuchi, L. Gradon, S. Suvachittanont and W.W. Szymanski, *Aerosol and Air Quality Research*, 6(1), 67-81, 2006.

[9] G. C. England et al, *Journal of the Air and Waste Management Association*, 57, 65-78, 2007.

[10] S. Win Lee, T. Herage, I. He and B. Young, *Powder Technology*, 180(1-2), 145-150, 2008.

[11] JIS Z 8901-1995: Test powders and test particles.

Reduction technology of carbon dioxide emission from a coal utilized power generation system

H. Makino & N. Noda
Central Research Institute of Electric Power Industry, Japan

Abstract

Coal is an important energy resource for meeting the further demand for electricity, as coal reserves are much more abundant than those of other fossil fuels. But, coal utilization technology produces carbon dioxide more than other fossil fuels, because of the higher carbon content of coal. For the control of the global warming problem, it is required to develop a new technology for the reduction of carbon dioxide emission from coal-utilized power stations.

To reduce carbon dioxide emission, it is very important to develop a new technology of high efficiency power generation system and utilize biomass energy of carbon neutral. For the further reduction of carbon dioxide, the removal and storage technology is an effective one. It is necessary to apply the smart process of powder technology for the development of coal utilized power generation system of new type including a high efficiency power generation system and carbon dioxide removal type power generation system. For the utilization of biomass energy, it is very important to develop the smart technology for the handling of biomass like drying, carbonizing, pulverizing and conversion technology.

In this paper, many technologies for the powder handling related to the high efficiency coal utilized power generation, biomass utilization and carbon dioxide removal power generation are introduced. Furthermore, the present level of these technologies and the development tendency for the future are surveyed.
Keywords: coal utilization, carbon dioxide, biomass, low rank coal, high efficiency power generation.

doi:10.2495/CHEM110091

1 Introduction

Coal is an important energy resource for meeting the further demand for electricity, as coal reserves are much more abundant than those of other fossil fuels. But, coal utilization technology produces carbon dioxide more than other fossil fuels, because of the higher carbon content of coal. For the control of the global warming problem, it is required to develop a new technology for the reduction of carbon dioxide emission from coal-utilized power stations.

To reduce carbon dioxide emission, it is very important to develop new technology of a high efficiency power generation system and utilize low carbonized coal or biomass energy of carbon neutral. Furthermore, carbon dioxide removal technology is one figure but it requires large power consumption and high cost.

In this paper, the utilization technology of low rank coal which is low carbon content, and the biomass utilization technology for the pulverized coal combustion are investigated at first. And, the development situation of the Integrated Coal Gasification Combined Cycle and Integrated Coal Gasification Fuel Cell Combined Cycle as a high efficiency power generation system is introduced. Finally, some methods of carbon dioxide removal technology and fuel upgrading technology are overviewed.

2 Utilization technology of sub-bituminous coal

As the low carbonized coal like sub-bituminous coal or lignite is lower carbon concentration than bituminous coal, the utilization of low carbonized coal has the possibility to reduce CO_2 emission. Especially, the minable reserve of sub-bituminous coal is over 30% of that of bituminous coal, so the utilization of sub-bituminous coal in pulverized coal combustion is investigated in Japan. But, the property of sub-bituminous coal is different from that of bituminous coal. For the reduction of CO_2 emission by using sub-bituminous coal, it is important to maintain the same power generation efficiency with bituminous coal. In this chapter, the new combustion technology for pulverized sub-bituminous coal is investigated in pulverized coal combustion test furnace in CRIEPI. The outline of this test furnace is indicated in Figure 1.

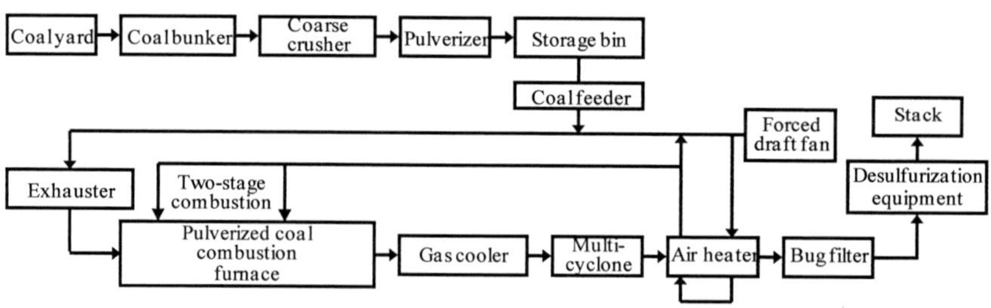

Figure 1: Outline of the coal combustion test facility.

2.1 Combustion technology of sub-bituminous coal

Figure 2 indicates the difference of the oxygen distribution in test furnace [1]. Fig. 2 (a) shows the distribution on bituminous coal combustion and Fig. 2 (b) shows that on sub-bituminous coal combustion. Oxygen consumption in furnace is affected by combustion situation, so this contour indicates the shape of combustion flame. The oxygen consumption rate near the burner exit on sub-bituminous coal combustion is lower than that on bituminous coal combustion, because the evaporation period of the moisture in sub-bituminous coal is necessary before combustion. So, the ignition of sub-bituminous coal is later compared with bituminous coal. And, after the ignition the flame on sub-bituminous coal combustion is diffused rapidly to the outer side of the furnace. This phenomenon is caused by the diffusion of pulverized coal particle. As sub-bituminous coal includes high concentration moisture, the coal particle of sub-bituminous coal becomes porous or bursts to fine particles by the rapid evaporation of moisture in the pore. Small coal particles or porous coal particles have weak inertia, so these particles are easy to be diffused by swirl force of combustion air.

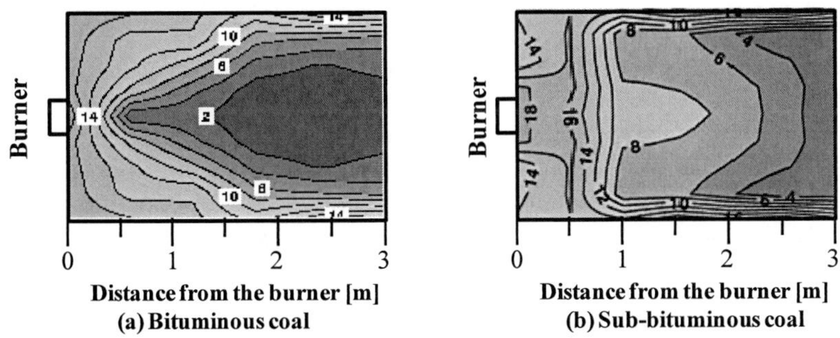

Figure 2: Difference of the oxygen distribution in test furnace.

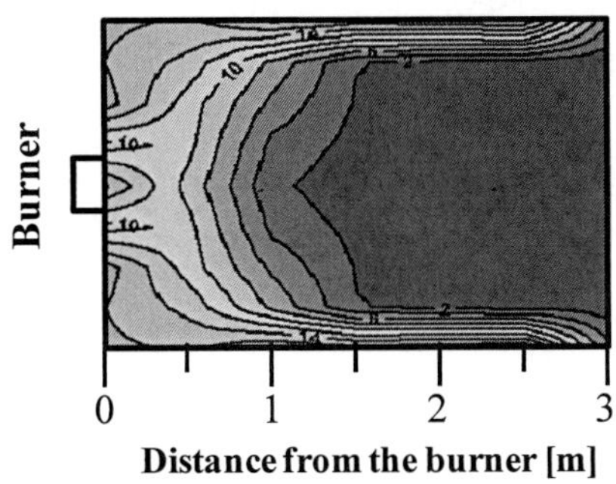

Figure 3: Oxygen concentration by modification on sub-bituminous coal.

As the solution method of this problem, the reduction of the flow rate of primary air and reduction of the swirl force of secondary air is effective. The reduction of primary air velocity can improve the ignition point closer to burner exit and the reduction of swirl force can reduce the diffusion of coal particle to the outer side of the furnace. Figure 3 indicates the oxygen concentration of the flame controlled by this modification on sub-bituminous coal combustion [1]. The flame shape of this condition is similar to the flame shape on bituminous coal combustion shown in Fig. 2 (a). In this condition, NOx and unburned carbon concentration in fly ash can be reduced same level with bituminous coal combustion and flame temperature distribution becomes similar of bituminous coal combustion, so the power generation efficiency is seemed to be same level with bituminous coal combustion.

2.2 Blend combustion of sub-bituminous coal with bituminous coal

In utility boiler, the coal whose property is different from usual coal is utilized by blend combustion with other kind coal. In this section, the blend combustion characteristics of sub-bituminous coal with bituminous coal are investigated.

Figure 4 indicates the NOx concentration and conversion to NOx of fuel bond nitrogen on blend combustion [2]. NOx concentration and conversion to NOx of fuel bond nitrogen on blend combustion become mean value of those of each coal combustion. Figure 5 indicates the unburned carbon concentration in fly ash and unburned fraction on blend combustion [2]. The unburned carbon concentration on blend combustion is higher than the mean value of those of each coal combustion. And, unburned carbon fraction on blend combustion has the maximum value at the condition of blend ratio of 25% sub-bituminous coal and 75% bituminous coal. This tendency is caused by the influence of moisture in sub-bituminous coal on bituminous coal combustion. The combustible in sub-bituminous coal is better than bituminous coal, so the unburned carbon fraction becomes low in sub-bituminous coal combustion. On the other hand, as the combustion characteristics of combustible in bituminous coal is worse than that of bituminous coal, the unburned carbon fraction of bituminous coal on blend combustion with sub-bituminous coal becomes higher by the influence of the moisture in sub-bituminous coal. To prevent this problem, it is effective to divide the injection burner for sub-bituminous coal and bituminous coal.

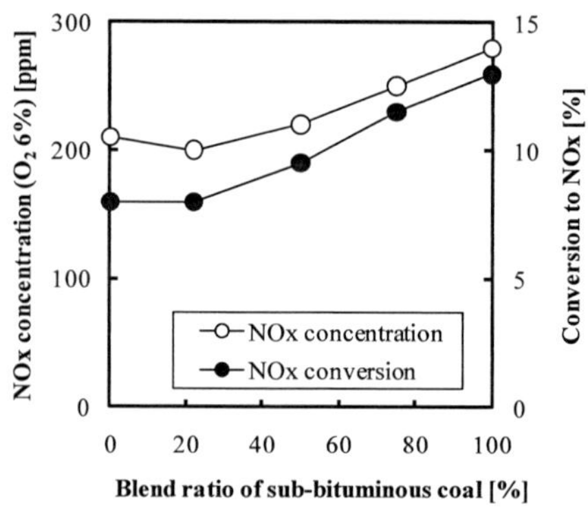

Figure 4: NOx concentration and conversion on blend combustion.

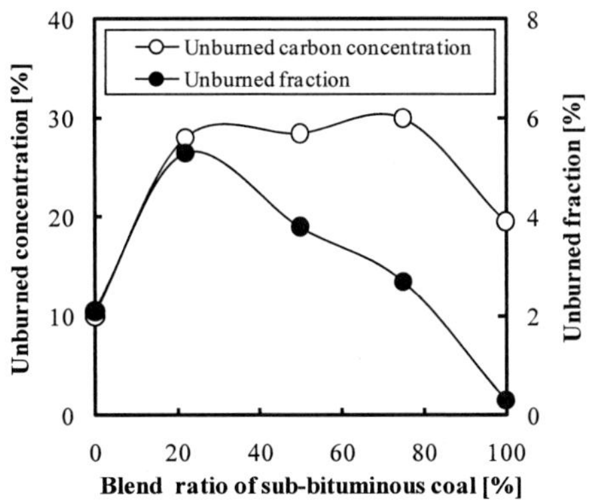

Figure 5: Unburned carbon concentration in fly ash and unburned fraction on blend combustion.

3 Utilization of biomass energy

Biomass is defined as carbon neutral energy because biomass is renewable. But, as there are many kinds of biomass species including woodchip, wood pellet, sewage sludge, and so on, the pulverizing characteristics and combustion characteristics of these species are different each other. At now, the usual method for the utilization of the biomass energy is blend combustion in pulverized coal combustion boiler. In CRIEPI, many kinds of biomass species are investigated on blend combustion with pulverized coal in test furnace.

Figure 6 indicates the conversion to NOx in pulverized coal combustion and blend combustion of dried sewage sludge with pulverized coal (Newlands coal) [3]. The conversion to NOx on blend combustion of dried sewage sludge with pulverized coal is lower than that of pulverized coal combustion. The dried sewage sludge includes higher volatile matter compared to Newlands coal and the fuel ratio is lower, so the formation of NOx in the flame is controlled by the effective formation of reduction flame. Figure 7 shows the unburned carbon concentration in fly ash in pulverized coal combustion and blend combustion of dried sewage sludge and pulverized coal (Newlands coal) [3]. As the fuel ratio of dried sewage sludge is lower than Newlands coal and the ash content of dried sewage is high, it is considered that the unburned carbon concentration on blend combustion becomes lower than that on Newlands coal combustion.

Although the biomass utilization in pulverized coal combustion boiler is effective method for CO_2 reduction, biomass worsens the pulverizing characteristics. So, it is necessary to develop the high performance pulverizing technology for biomass or mixture of coal and biomass. Furthermore, the potential usable volume of biomass is very low compared to the existed thermal power plant. So, it has to be noticed that the reduction effect of CO_2 emission by biomass utilization is not so much for the emission of CO_2 from thermal power plant.

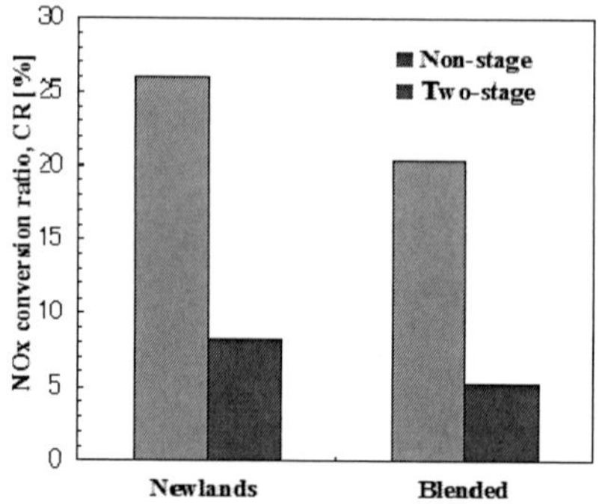

Figure 6: Unburned carbon concentration in fly ash and unburned fraction on blend combustion.

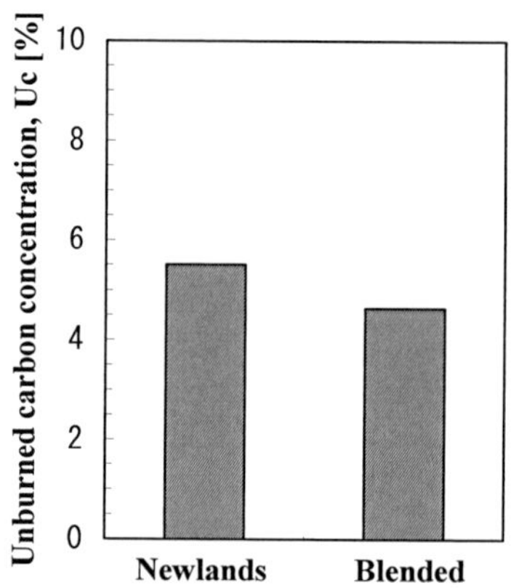

Figure 7: Unburned carbon concentration in fly ash in pulverized coal combustion and blend combustion of dried sewage sludge and pulverized coal.

4 Development of high efficiency power generation system

The high efficiency power generation system can reduce the CO_2 emission for the required electric power. In Japan, the ultra super critical (USC) boiler has been operated and the steam condition is improved step by step. On the other hand, for the great improvement of power generation efficiency, the development of the integrated coal gasification combined cycle (IGCC) and the integrated coal gasification and fuel cell combined cycle (IGFC) are promoted.

Figure 8 indicates the system flow of IGCC process. In Japan, the demonstration plant of IGCC has been operated since 2007. This plant is succeeded for continuous run of 5,000 hours. But, this plant uses a wet gas cleaning system. For the more improvement of power generation efficiency, the development of hot gas cleaning system and utilization of high temperature gas turbine are important. In hot gas cleaning system, the development of the high performance adsorption material for H_2S, HCl, HF and so on is required. And, the functional material for thermal barrier coating is required for the development of high temperature gas turbine. In the future, the improvement of the reliability and cost performance is required.

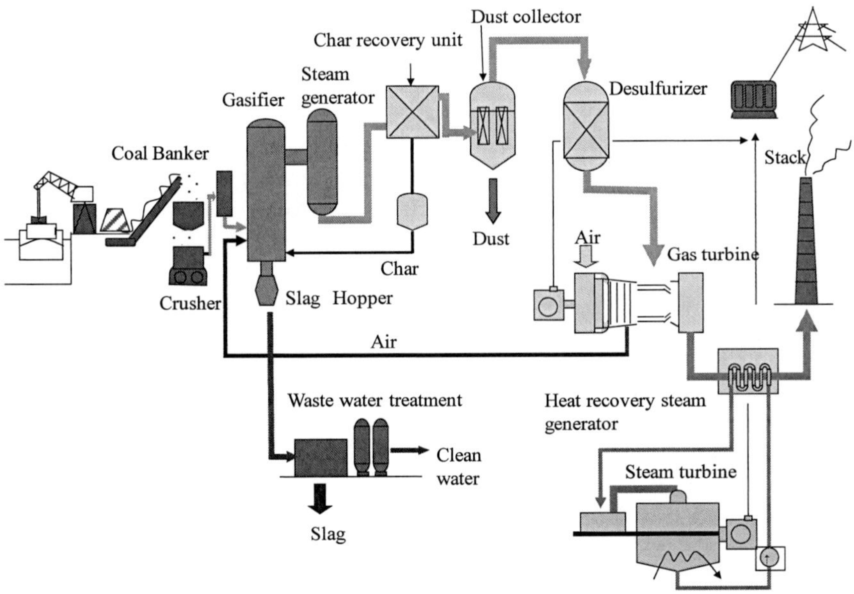

Figure 8: System flow of IGCC process.

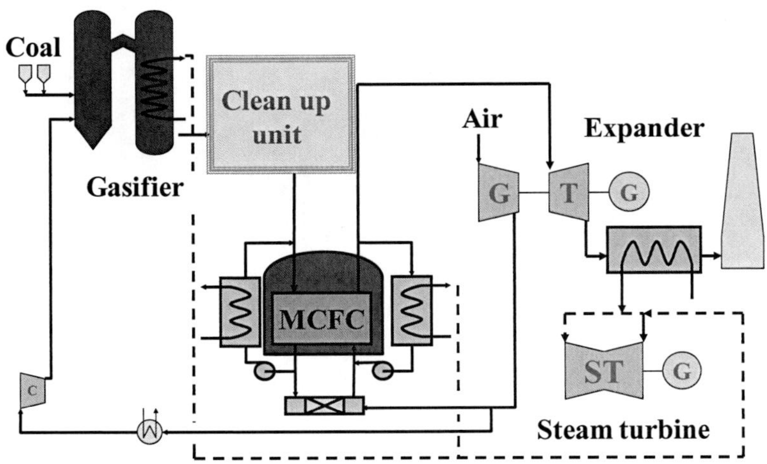

Figure 9: System flow of IGFC using MCFC.

The high temperature fuel cell can maintain high efficiency of power generation. IGFC can be reduced CO_2 emission more than IGCC. As a high temperature fuel cell, the molten carbonate fuel cell (MCFC) and the solid oxide fuel cell (SOFC) are investigated. Figure 9 shows the system flow of IGFC using MCFC. The main subjects of the utilization of MCFC are the improvement of the durability of the cell performance for long time and the cost performance. The cost of MCFC can be reduced about half and the cell performance can be kept over 90% of the initial performance after 40,000 hours' continuous run by the research using new smart material in CRIEPI.

5 Investigation of CO_2 capture and storage method

To reduce CO_2 emission greatly, the CO_2 capture and storage (CCS) system is one method. But, CO_2 emission from the thermal power plant is large amount, so CO_2 removal method requires the reduction of power generation efficiency and the increase of power generation cost extremely. Figure 10 indicates the comparison of estimation result of power generation efficiency with and without CO_2 removal. In general, the power generation efficiency of the thermal power plant with CO_2 removal is about 30% lower compared to that of the power generation plant without CO_2 removal. In these CO_2 removal systems, the CO_2 removal technology from IGCC system is better than the CO_2 removal technology from the pulverized coal combustion power plant. For more improvement of power generation efficiency of IGCC system with CO_2 removal, the development of high performance CO_2 separation technology including the absorption process, the adsorption process, membrane separation process and so on. It seems that CO_2 removal technology can be improved by the application of new smart material.

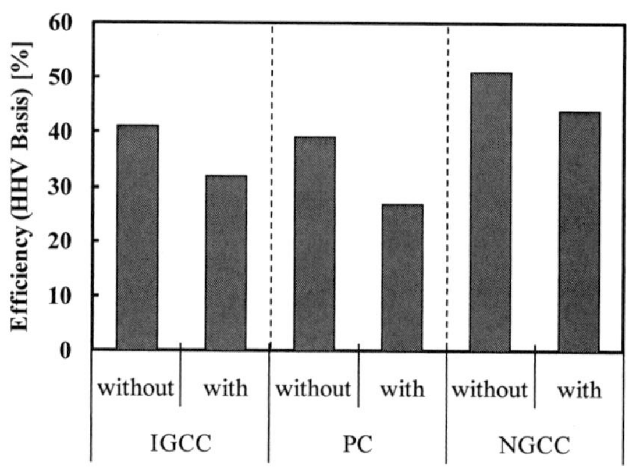

Figure 10: Comparison of estimation result of power generation efficiency with and without CO_2 removal.

In CRIEPI, to keep the high efficiency power generation of IGCC when CO_2 is removed, new type CO_2 removal IGCC system using O_2-CO_2 blown gasifier and O_2- CO_2 blown gas turbine. Figure 11 shows the flow of this system [4]. This system is supposed a combination of IGCC and OxyFuel (O_2- CO_2 blown) pulverized coal combustion system. So, in this system, gasifier and gas turbine are operated O_2 from the air separation unit and CO_2 from recycled flue gas. As CO_2 has gasification effect different from N_2 and the performance of CO_2 in gas turbine is higher than N_2, the power generation efficiency of this system becomes higher than the other system. The development of this system is progressed in bench scale plant level.

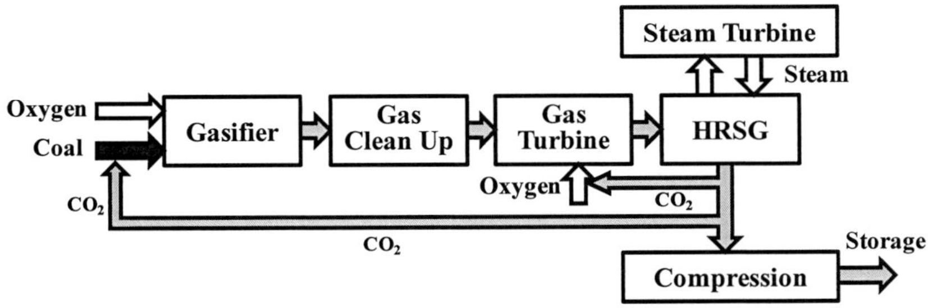

Figure 11: System flow of IGCC with CO_2 removal.

6 Upgrading of low rank coal and biomass

In chapter 2 and 3, the utilization technology of the low rank coal and biomass is introduced. For the utilization of the low rank coal and biomass, the big problem is the included moisture. Although many kinds of upgrading technology of low rank coal and biomass are developed, the drying technology is one of most important methods. But the moisture in low rank coal and biomass is not only bulk moisture but also moisture in the pore, so it is difficult to remove the moisture efficiently. For the high performance of moisture removal, the methods using evaporation are effective. On the other hand, these methods require the higher power consumption for the increase of coal temperature.

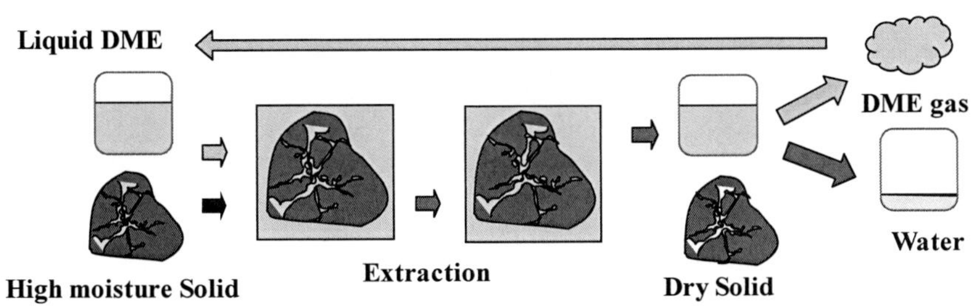

Figure 12: Concept of the extraction method using DME.

In CRIEPI, the extraction method using DME (di-methyl-ether) is investigated. Figure 12 shows the concept of this method [5]. DME is gas phase in ambient pressure and temperature, but it becomes liquid phase by pressurization in ambient temperature. And the liquefied DME can include moisture. So, in this system, the liquefied DME is mixed with the low rank coal or biomass, and DME extracts moisture. After the separation of liquefied DME including moisture with dried coal or biomass, DME converted to gas phase by decompression. And gaseous DME is separated with liquid water. The gaseous DME can be converted to liquid phase by pressurization, and DME is used again by recycling. This method can reduce the power consumption required for the drying of low rank coal and biomass.

7 Conclusion

Coal is an important energy resource for meeting the further demand for electricity, as coal reserves are much more abundant than those of other fossil fuels. For the coal utilization in the future, it is necessary to reduce CO_2 emission. To reduce the CO_2 emission, it is very important to develop new technology of high efficiency power generation system and utilize low carbonized coal and biomass energy of carbon neutral. For the further reduction of CO_2 emission, the CO_2 removal technology is effective one, but it requires high power consumption and cost.

It is necessary to apply the smart technology for new function material design and powder handing technology for the development of new type coal utilized power generation system including high efficiency power generation system and carbon dioxide removal power generation system. For the utilization of low rank coal and biomass energy, it is very important to develop the smart technology for the handling of low rank coal and biomass like drying, pulverizing and combustion. We'll continue to develop these technologies.

References

[1] Ikeda M., H. Makino, Y. Kozai, Emission Characteristics of NOx and Unburned Carbon in Fly Ash of Sub-bituminous Coal Combustion. *JSME International Journal*, Series B, 15, 3, pp506-511, 2002

[2] Ikeda M., H. Makino, H. Morinaga, K. Higashiyama, Y. Kozai, Emission Characteristics of NOx and Unburned Carbon in Fly Ash during Combustion of Blends of Bituminous/Sub-bituminous Coal Combustion. *Fuel*, 82, 13, pp1851-1857, 2003

[3] Tsuji H., H. Makino, H. Shirai, N. Ikemoto, K. Murai, S. Momoda, Blended Combustion of Dried Sludge Pellet with Pulverized Coal. *Journal of the Japan Institute of Energy*, 85, 9, pp769-772, 2006

[4] Oki Y., J. Inumaru, S. Hara, M. Kobayashi, H. Watanabe, S. Umemoto, H. Makino, Development of oxy-fuel IGCC system with CO_2 recirculation for CO_2 capture. *Proc. of the 10th Int. Conf. On Greenhouse Gas Control Technology*, 2010

[5] Kanda H., H. Makino, M. Miyahara, Energy-Saving Drying Technology for Porous Media using Liquefied DME Gas. *Adsorption*, 14, 2, 2008

Catalyst design with porous functional structures

J. Van Noyen, S. Mullens, F. Snijkers & J. Luyten
Flemish Institute of Technological Research (VITO),
Department of Materials Technology, Belgium

Abstract

The chemical industry is experiencing important changes. The driving force for these changes is a growing need to improve competitiveness and consolidate market positions while complying with the regulations for safeguarding human health, and the environment. Currently, what is known as "green chemistry" or "sustainable technology" is at the heart of the changes the chemical industry is undergoing.

Catalysis and materials science will play an important role in this new approach. One domain with specific relevance is porous ceramics and metals, substrates with pores sizes ranging from vacancies at the atomic level to macro pores with sizes of millimeters. There are plenty of emerging applications for porous functional components. Each application will specify the window of properties of the porous material.

A technology assessment of the developed porous materials and powder processing techniques for their use as catalyst or membrane system will be presented, limited to inorganic porous materials which can be synthesized by dry and wet powder processing methods. The applications overview for porous materials is focused on macroporous components with a designed functional coating.
Keywords: porous materials, powder processing techniques, catalysis, structured reactors, catalytic membrane reactors, coating techniques.

1 Introduction

It is widely acknowledged that there is a growing need for more environmentally acceptable processes in the chemical industry. This trend towards what has become known as "green chemistry" or "sustainable technology" necessitates a

WIT Transactions on Ecology and the Environment, Vol 154, © 2011 WIT Press
www.witpress.com, ISSN 1743-3541 (on-line)
doi:10.2495/CHEM110101

shift to new concepts that minimize waste at source, avoid the use of toxic and/or hazardous substances and are energy-efficient. To realize these needs specific tools have to be developed. Catalysis is generally accepted as a strong pillar within the green chemistry approach. Ceramic catalysts and catalyst supports constitute the largest market segment in high tech ceramics industry outside electronic ceramics. In 2005 the market was estimated $ 2.03 billion, with an expected increase of 5% per year to $ 3.16 billion in 2009 [1].

Catalysts are the key for sustainable development in the chemical process industry. They aid the synthesis of products in a resource protective way, with less consumption of energy, and in some cases, without any formation of by-products or waste. Catalysts, especially applied in a structured way, play an important role in the so-called integrated approach to environmental protection, which, among others, includes integration of various process operations such as chemical reaction, separation, heat exchange, and momentum transfer. The result of process intensification is the reduction of investment costs, which is often combined with significant energy recovery and space saving. Besides others, such savings are possible with the use of multifunctional reactors (such as monolith or membrane reactors, catalytic filters), reactive distillation columns, etc. [2].

2 Structured catalysts vs packed beds

Heterogeneous catalytic reactions account for over 85% of industrial chemical processes. They are involved in a large variety of processes ranging from refining over fine and specialty chemistry to environmental protection and sustainable chemistry [3]. Several processing and coating routes and related characterization techniques have been developed in order to manage the various application domains and window of properties. Classical heterogeneous catalysts consist of an active component, a binder, and several additives. These mixtures can be shaped in several ways, for instance by pressing, extrusion or spray drying. Depending on the shaping technique, wet or dry processing is required. Several shapes can be obtained, such as spheres, cylinders, tablets, lobed structures,… with dimensions of tens of μms up to several millimeters. These catalyst bodies can be used in fixed bed reactors, or in fluidized bed reactors. Several reviews on classical catalytic processes and the selection of a proper catalyst system can be found in [4–6].

Besides the well established classical processes with random fixed or fluidized beds, structured reactors have been developed in both chemical and automotive industries since the early 1970s. Structured internals can play a very important role and allow solutions that were previously impossible [7]. They allow a unique way of achieving process intensification in the chemical process- and refining industry. Several functions or processes are designed to occur simultaneously in multifunctional reactors. The aim is an optimal integration of mass, heat and momentum transfer within a single reactor vessel [8]. A structured reactor contains a macro-structured internal which can be made of ceramics, metal or carbon, situated inside the reactor. It can be considered as an intensified form of a packed bed reactor [9]. The advantage of such a structured reactor is that it may be designed in full detail up to the local surroundings of the catalyst, allowing ultimate precision. In addition, structured reactors are flexible with respect to different length scales, i.e., diffusion lengths, voidage,... Different manufacturing routes have been developed to produce such macroporous support structures.

Specific surface area and catalytic activity are obviously of crucial importance for obtaining a good catalyst with high activity and stability. Functionalization of the structured reactor internal is therefore an essential step. Reviews on the manufacture of macroporous supports can be found in [7, 10]. An intrinsic difficulty associated with the adoption of structured catalysts in reactors for chemicals production is the limited volume fraction of catalytically active material as compared to packed beds of catalyst pellets (structured catalyst ~20% compared to a packed bed ~60%). In this respect, however, one should consider that the effectiveness factors of the thin catalytic washcoats in structured catalysts are generally greater than those of pellets, and typically close to one.

To achieve comparable catalyst loadings with that of a randomly packed bed, either a high geometrical area of the structured internal or an integral structured catalyst is a prerequisite. If the catalyst is coated on the surface of the structured internal, then the stability and resistance of the coating to adverse conditions such as high temperature, temperature shocks, and high pressures needs to be considered. To enhance the catalytic activity the support, i.e. from an innovative support for catalyst to a true catalytic system, advanced surface engineering is applied: phase inversion, etching of the support surface, coating with zeolites and/or composites with catalytically-active compounds (transition metals, defect sites,...) applying sol gel, co-precipitation, electrolytical methods, texturing the surface with lithography,... Different methods to apply a catalytic layer onto a support have been reviewed [10, 11].

Packed beds show good activity due to high catalyst loading and longer residence times. However, packed beds often show incomplete catalyst wetting and poor mass transfer rates, due to undesirable effects in the fluid dynamics [9]. This can for example lead to hot spots and catalyst deactivation. Pressure drop over packed beds is generally high. Numerous products are commercialized as porous beads, which can be loaded in different reactor designs. The catalyst can be produced relatively cheap [12]. Despite efforts to ameliorate the mechanical properties (based on nanopowders, the addition of a second phase, doping or novel sintering techniques), problems like attrition remain key in packed bed technology. In view of the rapid improvements in the area of catalysis, leading to highly active catalyst particles, the issues discussed above will only become more pressing. Major challenges can be foreseen in the more classical active bed designs.

Monolithic catalysts take advantage of their favourable pressure behavior; and often find application in automotive applications and off-gas treatment. Both for gas-solid as gas-liquid reactions they can offer significant advantages compared to more traditional multiphase reactors. Low pressure drop compared to solid packed beds is the major advantage. Processes which require conversion of large volumes of gas mixtures, such as hydrogenations, the lower pressure drop over monoliths can result in lower investment- and working costs for the compressor. In comparison with slurry reactors the separation of the catalyst is avoided. Because in monolithic catalysts very thin layers of catalyst can be applied, an increase in selectivity by the avoiding of detrimental sequential reactions is possible. For fast reactions the costs for the expensive noble metal catalysts can remain limited. However, for very exothermal reactions efficient heat transfer can become a problem (e.g. hot spots), which can limit the use of such reactor systems [7, 9, 13].

Foam catalyst structures can also be used for very similar reaction types as monolithic catalysts, as many of the advantages of the honeycomb type reactors also apply for this type of catalyst supports. On top of that, good mixing is assured by the tortuous path the reactants have to follow through the packing [14]. However, design flexibility is low because of the random structure of foam.

Within the field of **structured packings**, KATAPAK®, a structured packing developed by Sulzer Chemtech is generally considered as state-of-the-art technology. This packing is often used in catalytic distillation, where reaction and separation are carried out in one process. Depending on design and support material used, mass- and heat transfer properties are generally good. For catalytic applications however a coating or catalyst section still needs to be applied. These types of packings can be used for reactive distillations, or as total oxidation catalyst carriers. [9, 15].

The **three-dimensional fiber deposition** (3DFD) of a ceramic or metallic structure is a new and innovative support preparation technique (also known as Robocasting). It combines the major advantages of the packings listed above, such as low pressure drop, good mass- and heat transfer, good mixing. On top of that, the technique allows making a catalyst structure by design rather than chance, in this way allowing very flexible and efficient use of catalyst in the reactor volume.

This manufacturing technique comprises the extrusion of a highly viscous metallic or ceramic paste through a tin nozzle, mounted on a CNC machine or x,y,z-table. The porous architecture is built layer-by-layer, as is shown in figure 1 [10, 16].

The major pros and cons of the different techniques are listed in table 1.

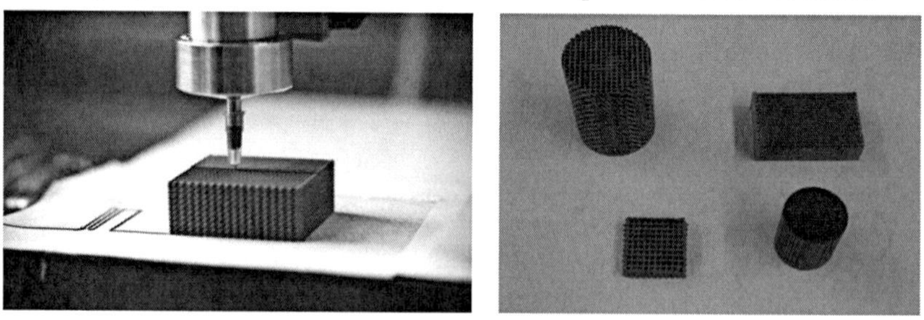

Figure 1: [Left] apparatus for three-dimensional fiber deposition; [Right] examples of parts built by 3DFD.

Table 1: Overview of different catalyst shaping techniques in view of relevant parameters for catalytic applications.

Parameter	Packed bed	Monolith	Foam	Structured packing	3DFD
Radial mass flow	fair	good	good	good	good
Radial heat exchange	fair	no	good	good	good
Tortuosity of fluid flow	yes	no	yes	yes	yes
Pressure drop	high	low	low	low	low
Geometrical macroporosity	35-40%	70-90%	60-90%	n.a.	up to 90%
Design flexibility	low	low	low	medium	high

Table 2: Overview of selected applications of structured catalysts.

Reaction	Support	Characteristics	Reference
catalytic combustion	ceramic foam	hydrocarbons preformed shapes	[17]
partial oxidation	ceramic foam and honeycomb	hydrocarbons selective low contact times	[18]
selective hydrogenation	honeycomb	cordierite monolith coated with Pd impregnated alumina	[19]
automotive exhaust	ceramic foam and honeycomb	three way catalyst, deNOx, soot filter	[20]
ethylbenzene to styrene	honeycomb	improved heat- and mass transfer	[21]
Fischer-Tropsch	honeycomb	coated cordierite and metallic honeycomb	[22]
anthraquinone autoxidation	honeycomb	reinforced amorphous silica, Pd catalyst	[23]
1-propanol + propionic acid to propyl propionate	Katapak®	reactive distillation, amberlyst 46 catalyst	[24]
steam reforming	ceramic foam	foam particles, pressure drop down 25%, heat transfer up 10%	[25]
ammonia oxidation	ceramic foam	800-1100°C selective no hot spots less Pt	[26]

A non exhaustive overview of catalytic reactions where structured catalysts have been used is given in Table 2. Further reviews and examples can be found [1, 4–7]. The application of structured packings for reactions with strong diffusion limitations, pressure drop constraints and heat transfer limitations will be the subject of intensified research in the next few years because of their obvious benefits of random (packed) beds.

3 Ceramic membranes

Ceramic membranes have a high thermal, chemical and mechanical resistance compared to polymeric membranes. These advantages can outweigh their higher intrinsic costs for certain applications, ensuring their share in the future growth of membrane technology [10, 27]. Membrane reactors integrate conversion and separation in one process step, thus allowing significant process intensification. The integration of membranes in a (catalytic) reactor allows to control the addition and distribution of one reactant in a controlled fashion in order to achieve optimal concentration profiles resulting in higher yields and achieve improved temperature control and safety. Alternatively, one of the products can be selectively removed from a process, typically used to shift equilibria and get round thermodynamics [28]. The concept of membrane reactors dates back to the 1960s, and since that time a large number of patents and papers have been published on that subject, at the intersection between catalysis, materials science and chemical engineering. The interest in membrane technology has been largely demonstrated at laboratory scale, i.e. for oxygen separation, hydrogenation, dehydrogenation, decomposition and oxidative reactions including partial oxidation and oxidative coupling of methane [28, 29]. Though at present only small industrial applications exist, the concept has yet a large opportunity to find widespread industrial application. Commercialization of such devices however, requires achieving the often incompatible aims of high performance, chemical stability, as well as optimal stacking, cost, sealing concepts, etc.

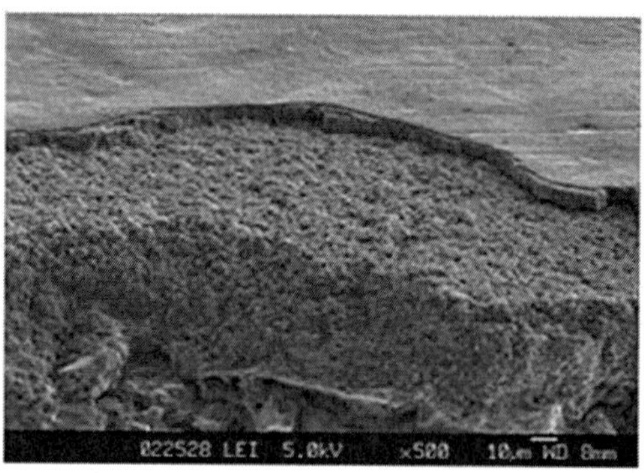

Figure 2: A multilayer ceramic membrane showing the different layers.

Tubes, hollow fibers and plates are the most common shapes. The macroporous substrates with pores of 5-10μm and a porosity of 30-50% offer the mechanical strength of the membrane. Mostly, tubes are produced by extrusion, hollow fibers are produced by spinning, and plates by pressing or tape casting. On top of the support, a functional layer is deposited to separate components from process streams, as shown in figure 2. Optimization of membranes is directed towards a combination of high flux of components through the membrane, with a high selectivity and thus separation.

A separate category is the dense membranes, the proton- and oxygen conductors (see figure 3). The main transport mechanism is mobility of vacancies of H- or O-ions at elevated temperatures. This ensures a 100% selectivity of the membrane. Performance of these membranes can be increased with a modified surface layer, i.e. by increasing roughness or applying a catalyst onto the membrane surface. These gas separating membranes, which currently draw a lot of attention in the field of energy applications, can also be of benefit for chemical reactors, by making catalytic membrane reactors using hollow fiber technology [30-31].

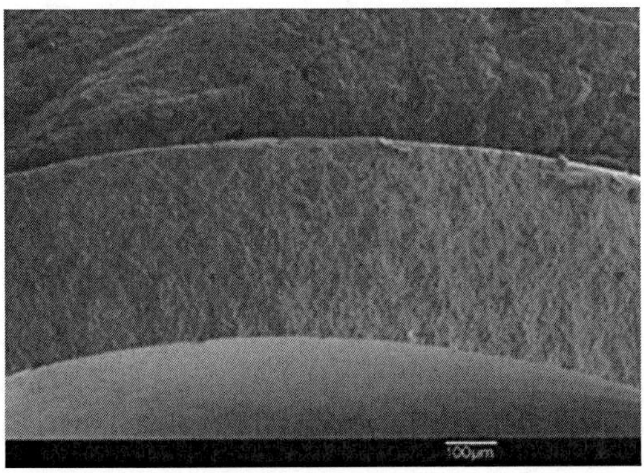

Figure 3: Dense oxygen separating hollow fibre membrane produced by a spinning technique.

4 Conclusion

There are plenty of emerging applications for porous components in different catalytic reactions. In order to cope with this large application domain and window of properties, several processing and coating routes and related characterization techniques have been developed enabling the manufacture of materials with a wide variety of porous architectures, pore size distribution (from a few angstroms to several millimeters), interconnectivity, pore gradients or layers with different pores sizes. This contribution has given an overview on the processing techniques which play a role in the design of inorganic materials for many applications.

Process intensification will be of key importance for the greening of our chemical industry, with optimization of heat- and mass transfer, and integration of separations as major fields for improvement.

Clearly, there is potential in ceramics and ceramic design techniques for catalysis and process intensification. Structured reactor internals allow for a flexible design of a catalytic structure, with flexibility down to the nanometer level by the functionalization of the surface. Porous functional structures, also taking ceramic membranes into account, allow integration of reaction and separation. For some applications they will certainly overcome the disadvantage of lower catalyst hold-ups and higher costs over their obvious advantages of higher reaction rates selectivities, lower pressure drop and better heat transport when compared to the traditional packed bed reactors.

References

[1] Keane, M.A., Ceramics for catalysis. *Journal of Materials Science*, **38**, pp. 4461-4675, 2003.

[2] Buchholz S., Future manufacturing approaches in the chemical and pharmaceutical industry. *Chemical Engineering and Processing*, **49(10)**, pp. 993-995, 2010.

[3] Centi, G. & Perathoner S., Integrated Design for Solid Catalysts in Multiphase Reactions. *Cattech*, **7(3)**, 78-89, 2003.

[4] Le Page, J.F., *Applied Heterogeneous Catalysis*, Editions Technip: Paris, pp. 1-286, 1987.

[5] Rase, H.F., *Handbook of Commercial Catalysts*, CRC Press: London, New York and Washington, pp. 1-482, 2000.

[6] Deutschmann, O., Knözinger, H., Kochloefl K. & Turek T., Heterogeneous Catalysis and Solid Catalysts. *Ullmann's Encyclopedia of Industrial Chemistry*, Wiley-VCH Verlag GmbH: Weinheim, Electronic version, 2009.

[7] Scheffler, F., Claus, P., Schimpf, S., Lucas, M. & Scheffler, M., Heterogeneously Catalyzed Processes with Porous Cellular Ceramic Monoliths (Chapter 5.4). *Cellular Ceramics: Structure, Manufacturing, Properties and Applications*, ed. M. Scheffler & P. Colombo, Wiley-VCH GmbH: Weinheim, pp. 454-483, 2005.

[8] Dautzenberg, F.M. & Mukherjee, M., Process intensification using multifunctional reactors. *Chemical Engineering Science*, **56**, pp. 251-267, 2001.

[9] Pangarkar, K., Schildhauer, T.J., van Ommen J.R., Nijenhuis, J., Kapteijn, F. & Moulijn J.A., Structured packings for multiphase catalytic reactors. *Industrial & Engineering Chemical Research,* **47**, pp. 3720-3751, 2008.

[10] Luyten, J., Mullens S. & Thijs, I., Designing with pores – Synthesis and Applications. *KONA Powder and Particle Journal*, **28**, pp. 1-12, 2010.

[11] Meille, V., Review on methods to deposit catalysts on structured surfaces. *Applied Catalysis A: General*, **315**, pp. 1-17, 2006.

[12] Holt, E.M., The properties and forming of catalysts and adsorbents by granulation. *Powder Technology*, **140**, pp. 194-202, 2004.
[13] Tomasic, V & Jovic, F., State-of-the-art in the monolithic catalysts/reactors. *Applied Catalysis A: General*, **311**, pp. 112-121, 2006.
[14] Richardson, J.T., Demue, D. & Hung, J.-K., Properties of ceramic foam catalyst supports: mass and heat transfer. *Applied Catalysis A: General*, **250**, pp. 319-329, 2003.
[15] Pfeuffer, B., Kunz, U., Hofmann, U., Turek, T. & Hoell D., Heterogeneous reactive extraction for an intensified alcohol process. *Catalysis Today*, **147S**, pp. S357-S361, 2009.
[16] Nuyts, P., Luyten, J., Mullens, S., Schroeven, M. & Rombouts, M., The 3D fiber deposition technique for producing porous periodic metals and ceramics. *Proceed. ISuP - 2010 conference*, Bruges, pp. 1-4, 2010.
[17] Richardson, J.T., Peng, Y. & Remue, D., Properties of ceramic foam catalyst supports: pressure drop. *Applied Catalysis A: General*, **204**, pp. 19-32, 2000.
[18] Torniainen, P.M., Chu, X. & Schmidt, L.D., Comparison of monolith-supported metals for the direct oxidation of methane to syngas. *Journal of Catalysis*, **146(1)**, pp. 1-10, 1994.
[19] Smits, H.A., Stankiewicz, A., Glasz, W.C., Fogl, T.H.A. & Moulijn, J.A., Selective three-phase hydrogenation of unsaturated hydrocarbons in a monolithic reactor. *Chemical Engineering Science*, **51(11)**, pp. 3019-3025, 1996.
[20] Li, L., Zhang, F. & Guan, N., Ir/ZSM-5/cordierite monolith for catalytic NOx reduction from automobile exhaust. *Catalysis Communications*, **9(3)**, pp. 409-415, 2008.
[21] Liu, W., Addiego, W.P. & Sorensen, C.M., Monolith Reactor for the Dehydrogenation of Ethylbenzene to Styrene. *Industrial & Engineering Chemistry Research*, **41(13)**, pp. 3131-3138, 2002.
[22] Visconti, C.G., Tronconi, E., Lietti, L., Groppi, G., Forzatti, P., Cristiani, C., Zennaro, R. & Rossini S., An experimental investigation of Fischer–Tropsch synthesis over washcoated metallic structured supports. *Applied Catalysis A: General*, **370(1-2)**, pp. 93-101, 2009.
[23] Edvinsson Albers, R., Nyström, M., Siverström, M., Sellin, A., Dellve, A.-C., Andersson, U., Herrmann, W. & Berglin, T., Development of a monolith-based process for H2O2 production: from idea to large-scale implementation. *Catalysis Today*, **69(1-4)**, pp. 247-252, 2001.
[24] Kotora, M., Buchaly, C., Kreis, P., Górak, A. & Markoš, J., Reactive distillation — experimental data for propyl propionate synthesis. *Chemical Papers*, **62(1)**, pp. 65-69, 2008.
[25] Twigg, M.V. & Richardson, J.T., Theory and Applications of Ceramic Foam Catalysts, *Chemical Engineering Research and Design*, **80(2)**, pp. 183-189, 2002.
[26] Campbell, L.E., Catalyst for the production of nitric acid by the oxidation of ammonia, *US005256387*, pp. 1-5, 1994.

[27] Buekenhoudt, A., Kovelvski, A., Luyten, J. & Snijkers, F., Basic aspects in inorganic membrane preparation (chapter 1.11). *Comprehensive Membrane Science and Engineering*, ed. Drioli, E. & Giorno, L., Elsevier: Amsterdam, pp. 217-252, 2010.

[28] Ravanchi, M.T., Kaghazchi T. & Kargari, A., Application of membrane separation processes in petrochemical industry: a review. *Desalination*, **235**, pp. 199-244, 2009.

[29] Caro, J., Schiestel, T., Werth, S., Wang, H. & Noack, M., Can inorganic membranes compete with organic ones? Perovskite hollow fibres for O_2-separation and support H_2-selective membranes. *Desalination*, **199**, pp. 365-367, 2006.

[30] Caro, J., Catalysis in micro-structured membrane reactors with nano-designed membranes. *Chinese Journal of Catalysis*, **29(11)**, pp. 1169-1177, (2008).

[31] Kovalevsky, A., Buysse, C., Snijkers, F., Buekenhoudt, A., Luyten, J., Kretzschmar, J. & Lenaerts, S., Oxygen exchange-limited transport and surface activation of Ba0.5Sr0.5Co0.8Fe0.2O3−δ capillary membranes. *Journal of Membrane Science*, **368(1-2)**, pp. 223-232, 2011.

Development of photonic and thermodynamic crystals conforming to sustainability conscious materials tectonics

S. Kirihara, N. Ohta, T. Niki, Y. Uehara & S. Tasaki
Joining and Welding Research Institute, Osaka University, Japan

Abstract

Photonic crystals with periodic variations in dielectric constants can exhibit band gaps in transmission spectra of electromagnetic waves through Bragg diffractions. Introduced structural defects can confine the electromagnetic energies at specific wavelengths, and localized modes of transmission peaks can be formed in above mentioned forbidden gaps. Moreover, thermodynamic crystals with periodically percolated morphologies of metals and ceramics can realize uniform dispersions in thermal and kinetic energies effectively. Introduced artificial dislocations or graded structures into the ideal periodicities realize heat and stress flow controls intentionally. These artificial crystals were processed by micro stereolithography. The electromagnetic wave transmittances, thermal conductivities and mechanical properties for the formed photonic and thermodynamic crystals were measured and compared with simulated results. The photonic and thermodynamic crystals will be applied to novel electromagnetic wave resonators and mechanical tools, respectively. Effective energy flow control with reducing losses will be realized along sustainability conscious strategies of artificial materials tectonics.
Keywords: photonic crystal, thermodynamic crystal, stereolithography.

1 Introduction

Photonic crystals with periodic arrangements of dielectric lattices can form forbidden gaps in electromagnetic waves spectra [1, 2]. These artificial crystals can totally reflect the light or electromagnetic waves at wavelengths comparable to the lattice spacing by Bragg deflection. The introduced structural defects in the periodic arrangements can localize the electromagnetic wave energies and

WIT Transactions on Ecology and the Environment, Vol 154, © 2011 WIT Press
www.witpress.com, ISSN 1743-3541 (on-line)
doi:10.2495/CHEM110111

form the transmission mode peaks in the band gaps according to the sizes and dielectric constants of the defects. Recently, we have newly developed micro stereolithography system to fabricate the micro photonic crystals composed of the ceramic lattices for the electromagnetic wave control in terahertz frequency ranges [3, 4]. In the near future, the terahertz wave will be expected to apply to various types of novel sensors which can detect gun powders, drugs, bacteria in foods, micro cracks in electric devices, cancer cells in human skin and other physical, chemical and living events [5, 6]. In this study, the ceramic lattices with diamond structures were created to realize the perfect band gap formations opening for all crystal directions [7, 8]. Subsequently, we successfully fabricated twinned diamond structures with plane defects between the mirror symmetric lattice patterns [9–11]. Localized modes to transmit the terahertz wave selectively were formed in the band gap region. Transmission spectra in the terahertz range through the alumina lattices were measured. The selective transmission modes in the band gaps were observed for the twinned diamond structures, and localized mode behaviours of the terahertz waves in the plane defects will be discussed.

Thermodynamic crystals with periodic arrangements of ceramic and metal phases were fabricated artificially to control heat and stress flows intentionally by computer aided design, manufacturing and evaluation [12]. Graded structures in volume fractions of ceramic and metal phases were created successfully to realize directional controls of heat and stress energies distributions. In this study, the alumina structures including air spheres with body centered cubic patterns were processed by using the micro stereolithography, and resin slurries with pure copper particles were infiltrated into the ceramics objects. Through dewaxing and sintering treatments, the thermodynamic crystals with the metals spheres arrangements in the ceramics matrices could be obtained exactly. Moreover, the functionally graded materials with continuous variations in volume fractions of the ceramics and metals were fabricated. The intermediate regions with the percolated morphologies were formed between the copper and alumina layers. The artificial microstructures composed of the alumina and pure copper phases were observed by using a digital optical microscope and a scanning electron microscope. The part accuracies of the periodic structures are measured. Thermal and mechanical properties were theoretically simulated method. The directional propagations of the heat and stress energy flows in the periodically percolated materials morphologies will be considered theoretically.

2 Experimental procedure

The photonic and thermodynamic crystals with the diamond lattice and body center cubic structure were spatially modelled by a computer graphic application (Think Design: Toyota Keram, Japan) as shown in Figures 1 and 2, respectively. In the case of diamond photonic crystal, the lattice constant was designed as 1 mm. The whole structure was $6\times6\times2$ mm^3 in size, consisting of 72 unit cells. The aspect ratio of the lattices was adjusted to be 1.5. In the case of thermodynamic crystal, the inverse model of $10\times10\times10$ mm in size included the

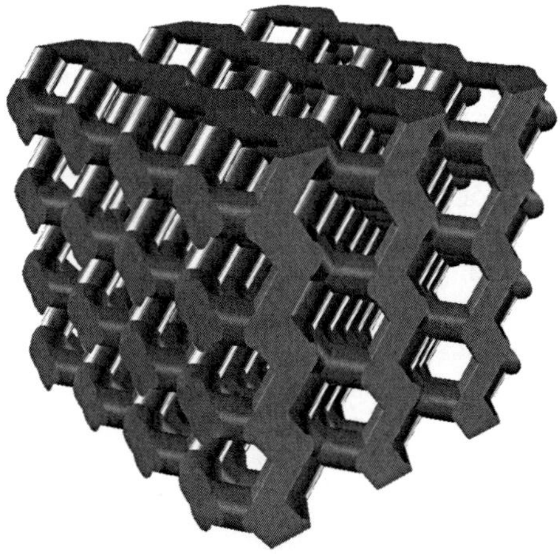

Figure 1: A computer graphic model of photonic crystal with diamond lattice arrangements to create periodic variations of dielectric constants.

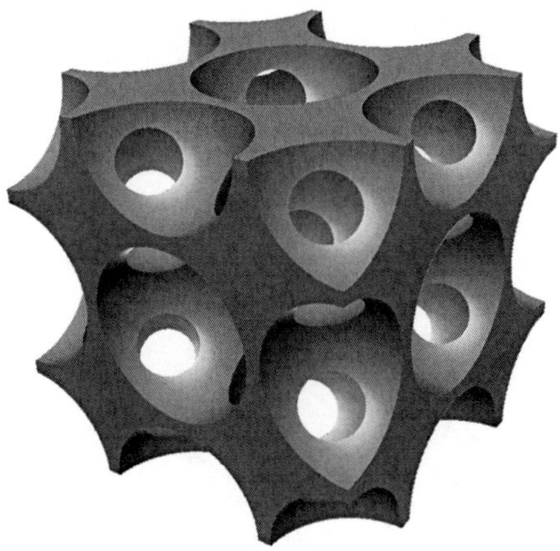

Figure 2: The computer graphic model of thermodynamic crystal with inverse sphere arrangements to create a body centered cubic structure.

air spheres of 4.0 mm in diameter. The lattice spacing for <100> direction was 4.5 mm. The volume ratio of the matrix was 26.5%. These designed models were converted into the numerical data of stereolithographic format and sliced into thin sections.

In the fabrication process of the photonic crystals, a micro pattering stereo-lithography system (SI-C1000: D-MEC, Japan) was used as shown in Figure 3. Photo sensitive acrylic resins dispersed with alumina particles of 170 nm in diameter at 40 volume % were fed over substrates from dispenser nozzles. The highly viscous resin paste was fed with controlled air pressure. It was spread uniformly by moving a knife edge. The thickness of each layer was controlled to

10 μm. A two dimensional pattern was formed by illuminating visible laser of 405 nm in wavelength on the resin paste surface. The high resolution has been achieved by using a digital micro mirror device (DMD) and an objective lens. Figure 3 shows a schematic of the micro stereolithography system. The DMD is an optical element assembled by micro mirrors of 14 μm in edge length. The tilting of each tiny mirror can be controlled according to the two dimensional cross sectional data transferred for a computer. The three dimensional structures were built through stacking these patterns. In order to avoid deformation and cracking during dewaxing, careful investigation for the heat treatment processes were required. The precursors with diamond structures were heated at various temperatures from 100 to 600°C while the heating rate was 1.0°C /min. The dewaxing process was observed in respect to the weight and color changes. The alumina particles could be sintered at 1500 or 1000°C, respectively. The heating rate was 8.0°C/min. The density of the sintered sample was measured by the Archimedes method. The microstructures were observed by optical microscope and scanning electron microscopy (SEM). The transmittance and the phase shift of incident terahertz waves were measured by using terahertz time domain spectroscopy (TDS) device (J-Spec2001: Aispec, Japan). The measured terahertz properties were compared with theoretical simulations by using a transmission line modeling (TLM) program (Micro-Stripes: Flomerics, EU).

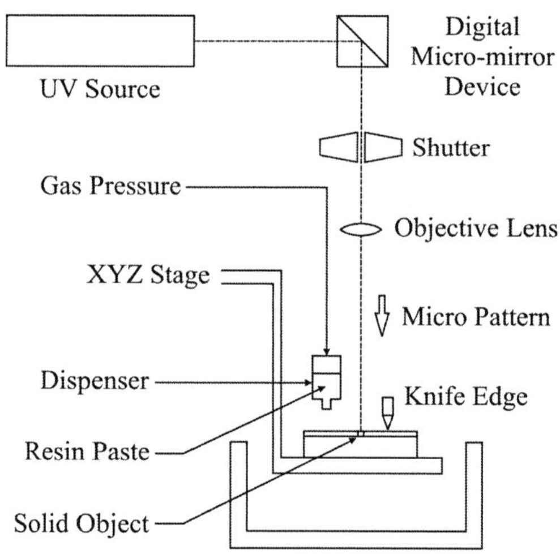

Figure 3: A schematic illustration of a micro pattering stereolithography by using continuous exposure techniques of ultra violet fine images.

The ceramics models of the inversed body centered cubic structures were fabricated automatically by utilizing a laser scanning stereolithographic machine (SCS-300P: D-MEC, Japan) as shown in Figure 4. The alumina particles of 1.8 μm in average diameter were dispersed into the photo sensitive acrylic resin at 70% in volume contents. The composite resin paste was spread on a grass substrate with 50 μm in layer thickness by the mechanical knife edge. An ultraviolet laser beam of 355 nm in wavelength and 100 μm in spot diameter was scanned on the resin surface to create two dimensional images with 10 μm in part accuracy. Through the layer stacking, sub-millimeter order three dimensional composite models were formed successfully. These obtained precursors were dewaxed and sintered at 600 and 1500°C for 2 hs with the heating rate of 1 and 8°C/min, respectively. Subsequently, the polyester resin past mixed with the pure copper particles of 75 μm in average diameter at 55% in volume contents were percolated into the inverse ceramic structures. After the dewaxing for the surly infiltrated samples at 800°C for 2 hs with heating rate of 1 °C/min, the pure copper particles were sintered at 1000 °C for 2hs with heating rate of 8°C/min in an argon atmosphere. The densities of the sintered samples were measured by Archimedes method. The part accuracy of the metal spheres arrangements in the ceramic matrices were measured by using the digital optical microscope (DOM) system (VHX-200: Keyence, Japan). The microstructures of ceramic and metal phases were closely observed by using the SEM. The heat and stress flows were simulated theoretically in the thermodynamic crystals by using a finite element method (FDTD) application (Ansys: Cybernet Systems, Japan).

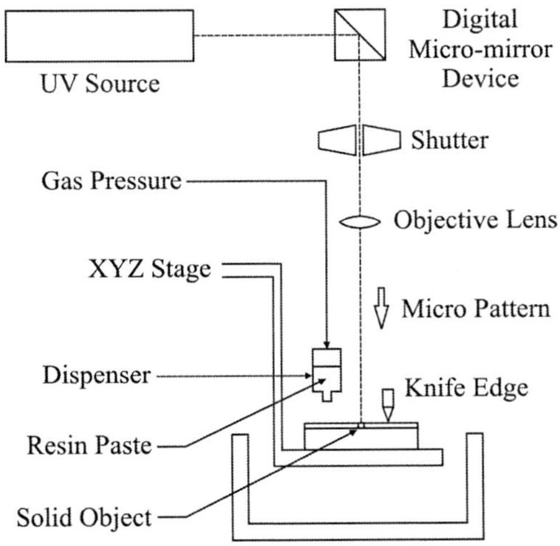

Figure 4: The schematic illustration of a stereolithography system by using ultra violet laser scanning techniques of cross sectional patterns.

3 Results and discussion

The lattice structures composed of the alumina dispersed acrylic resin were processed exactly by using the micro stereolithography. The spatial resolution was approximately 0.5%. Figure 5 shows a (100) plane of the sintered diamond structure composed of the micrometer order alumina lattice. The deformation and cracking were not observed. The linear shrinkage on the horizontal axis was

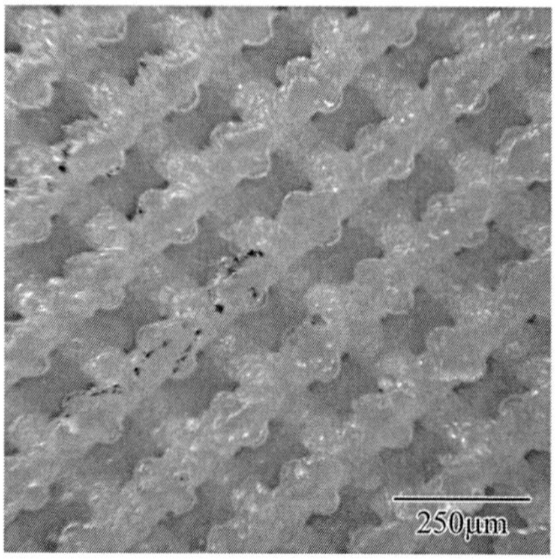

Figure 5: A micro photonic crystal composed of alumina ceramic lattices with the diamond structure fabricated by using the stereolithography.

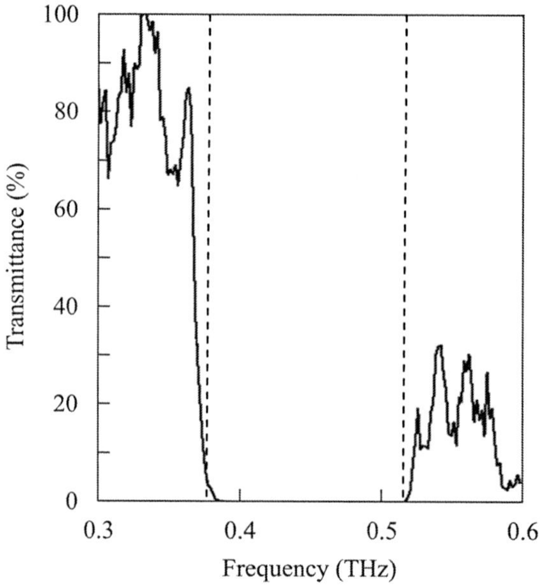

Figure 6: A measured transmission spectrum of terahertz waves for the Γ-X <100> direction in the micro alumina photonic crystal.

23.8% and that on the vertical axis was 24.6%. The relative density reached 97.5%. Dense alumina microstructure was formed, and the average grain size was approximately 2 μm. The measured dielectric constant of the lattice was about 10. The terahertz wave attenuation of the transmission amplitude through the alumina diamond structure for the Γ-X <100> crystal direction is shown in

Figure 6. The forbidden gap is formed at the frequency range of from 0.37 to 0.52 THz. The dotted lines show the higher and lower band gap edges calculated by using the TLM method. The similar transmission spectra for the Γ-K <110> and Γ-L <111> directions were obtained. These measured results of the photonic band gap frequencies were verified to have good agreements with the simulated

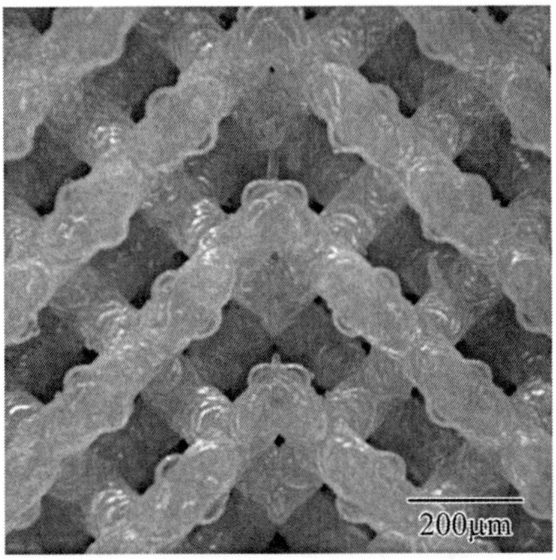

Figure 7: The twinned diamond photonic crystal with a plane defect between mirror symmetric lattices parallel to the (100) layer stacking.

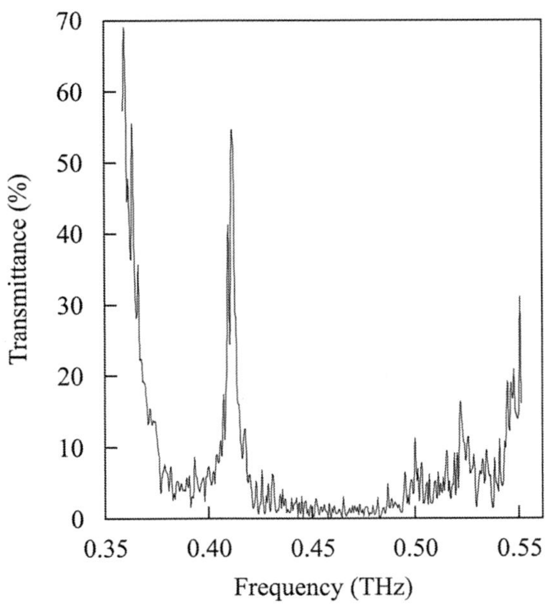

Figure 8: A localized mode formed in the photonic band gap range through the twinned diamond structure of the alumina lattices.

ones within a tolerance of 5%. A common region of the band gaps was observed in every crystal direction at the frequency range from 0.40 to 0.47 THz, where the electromagnetic wave cannot transmit through the lattices. Figure 7 shows the twinned diamond structure composed of mirror symmetric alumina lattices. The plane defect forms parallel to the (100) crystal layers. The transmission spectrum for the Γ-X <100> crystal direction of the twinned diamond structure is shown in Figure 8. The localized mode forms in the photonic band gap. At the peak frequency, the incident terahertz wave localized strongly in the plane defect, and the amplified wave propagated to the opposite side. The photonic

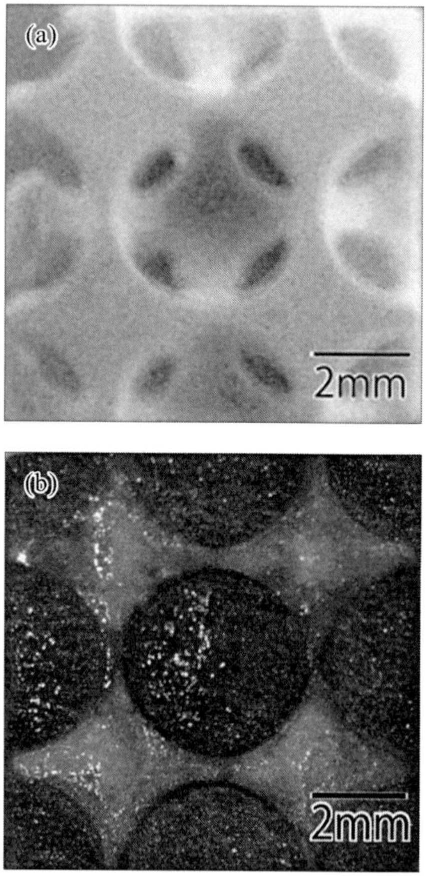

Figure 9: A sintered alumina lattice with the inverse body centered cubic structure fabricated by the stereolithography of a structural joining process (a) and a fabricated thermodynamic crystal with periodic arrangements of pure copper spheres in an alumina matrix through powder sintering treatments (b).

crystal with the twinned diamond structure to form the band gap and the localized mode will be applied to the effective terahertz wave antennas, resonators and filters.

The inverse models composed of the alumina dispersed acrylic resin were processed successfully by using the stereolithography. The spatial resolution was approximately 0.5% in size. Figure 9(a) shows (100) planes of the sintered body centered cubic structure composed of the sub-millimeter order alumina bulk including the air spheres. The deformation and cracking were not observed. The linear shrinkage on the horizontal axis was 6.5% and that on the vertical axis was 7.8%. It is possible to obtain the uniform shrinkage by designing an appropriate elongated structure in the vertical direction for compensation to the gravity effect. The measured relative density reached 98.5%. Dense alumina microstructure was formed, and the average grain size was approximately 2 μm. Figure 9(b) shows the (100) plane in the sintered thermodynamic crystal model composed of the pure copper spheres and alumina matrix. Diffraction peaks indicating the formation of copper oxide phases did not appeared in X-ray diffraction spectra. Cracks and pores as the sintering defects were not observed in the copper phase microstructures. The smooth interfaces between the copper spheres and alumina matrix were obtained successfully. These material phases were joined continuously without intermediate layers. Three dimensional heat and stress distributions in the fabricated thermodynamic crystals were simulated as shown in Figure 10(a) and (b), respectively. The heat flow transmitted from the source plane to the opposite side through the connected copper spheres of the percolated metal phases with high thermal conductivities. Compare with this, the tensile stress were distributed periodically into the alumina lattice parts of the distributed ceramics phases with high Young's modulus. Through introductions of point or plane defects in to the perfect periodic structures, heat and stress distributions can be localized into the specific regions to control the thermal and

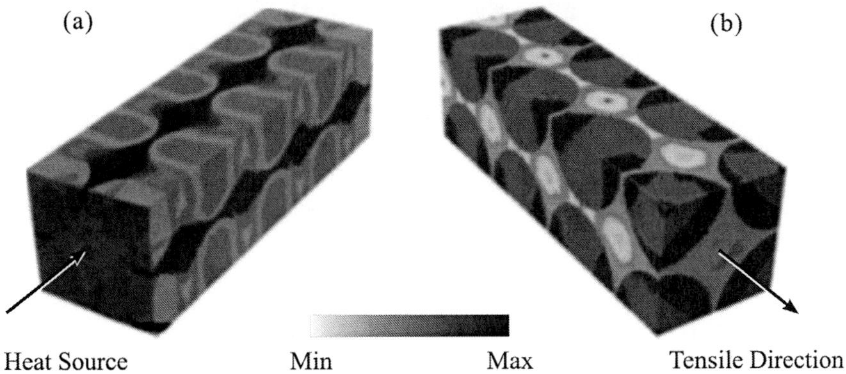

Figure 10: The computer graphics of simulated heat and stress distributions in the fabricated thermodynamic crystals with the body centered cubic structure. The graphics (a) and (b) show the heat and stress flows through the metal spheres and the ceramic matrix, respectively.

mechanical properties of the artificial crystals. Figure 11 shows the simulated result of the temperature distribution in the graded body centered cubic structure composed of the copper spheres with the varied diameters form 1.75 mm to 1.9 mm in the alumina matrix. The heat flow transmitted from the source plane to the opposite side can be curved for the metal spheres arrangements with the larger diameter along the structural gradient. The thermodynamic crystals with the graded structures are considered to be applied to the effective heating and cooling devices in the novel mechanical tools, internal combustion engines and thermoelectric conversion elements along the artificial material designing.

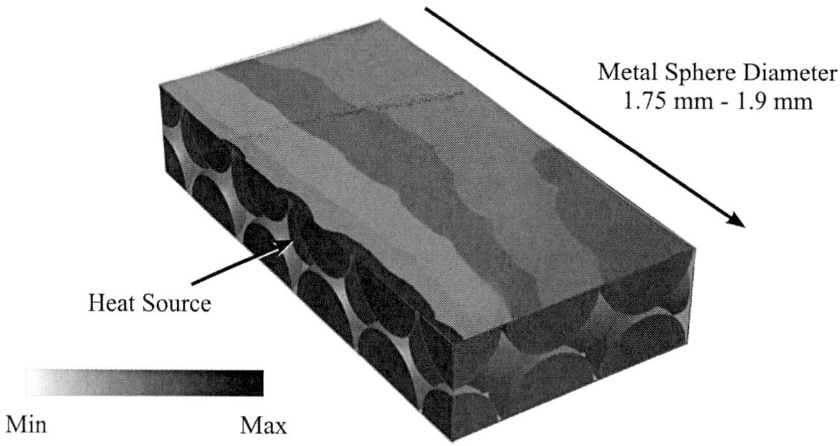

Figure 11: The simulated heat flow in the thermodynamic crystal of the graded structure with the body centered cubic pattern. The metal spheres with varied diameters were arranged in the ceramics matrix.

4 Conclusions

We have successfully fabricated photonic and thermodynamic crystals by using stereolithography and careful optimized dewaxing and sintering processes. The alumina photonic crystals with a diamond structure formed complete band gaps in terahertz wave frequencies. A twinned diamond crystals including with a plane defect between mirror symmetric lattice structures was also fabricated. A localized mode of a transmission peak was observed clearly in a forbidden band. In theoretical simulations of electromagnetic waves, localized mode profiles with multiple reflections were formed in a twinned interface. These micro dielectric components have strong potentials to be used as novel terahertz wave devices. Subsequently, the thermodynamic crystals with body centered cubic structures composed of pure copper spheres arrangements in alumina matrices were formed exactly. Heat and stress distributions in the crystals were simulated theoretically. The heat flows transmitted through the connected copper spheres with high thermal conductivities, and the tensile stresses were distributed periodically into the alumina lattices with high Young's modulus. Graded body centered cubic structures composed of the copper spheres with varied diameters in the alumina matrix were formed. The heat flow distributions could be curved along the graded direction for the copper spheres arrangements with the larger diameters. These artificial crystal structures are considered to be applied to the effective heating and cooling devices in the near future industrial fields.

Acknowledgements

This study was supported by Priority Assistance for the Formation of Worldwide Renowned Centers of Research - The Global COE Program (Project: Center of Excellence for Advanced Structural and Functional Materials Design) from the Ministry of Education, Culture, Sports, Science and Technology (MEXT), Japan.

References

[1] Ohtaka, K., Energy bang of photons and low energy photon diffraction. *Physical Review B*, **19(10)**, pp. 5057-5067, 1979.

[2] Yablonovitch & E., Inhabited spontaneous emission in solid-state physics and electronics. *Physical Review Letter*, **58(20)**, pp. 2059-2062, 1987.

[3] Chen, W., Kirihara, S. & Miyamoto, Y., Three-dimensional micro-photonic crystals of ZrO_2 toughened Al_2O_3 for terahertz wave applications. *Applied Physics Letter*, **91(15)**, pp. 153507-1-3, 2007.

[4] Miyamoto, Y., Kanaoka. H. & Kirihara, S., Terahertz wave localization at a three-dimensional ceramic fractal cavity in photonic crystals. *Journal of Applied Physics*, **103**, pp. 103106-1-5, 2008.

[5] Van Exter, M., Fattinger, C. & Grischkowsky, D., Terahertz time-domain spectroscopy of water vapour. *Optics Letters*, **14(20)**, pp. 1128-1130, 1989.

[6] Clery, D., Brainstorming their way to an imaging revolution. *Science*, **297**, pp. 761- 763, 2002.

[7] Ho, K., Chan, C. & Soukoulis, C., Existence of a photonic gap in periodic dielectric structures, *Physical Review Letter*, **65(25)**, pp. 3152-3165, 1990.

[8] Kirihara, S., Miyamoto, Y., Takenaga, K., Takeda, M. & Kajiyama, K., Fabrication of electromagnetic crystals with a complete diamond structure by stereolithography. *Solid State Communications*, **121(8)**, pp. 435-439, 2002.

[9] Kirihara, S. & Miyamoto, Y., Terahertz wave control by using ceramic photonic crystals with diamond structure including plane defects fabricated by micro-stereolithography. *The International Journal of Applied Ceramic Technology*, **6(1)**, pp. 41-44, 2009.

[10] Kirihara, S., Niki, T. & Kaneko, M., Three-dimensional material tectonics for electromagnetic wave control by using micro-stereolithography. *Ferroelectrics*, **387**, pp. 102-111, 2009.

[11] Kanaoka, H., Kirihara, S. & Miyamoto, Y., Terahertz wave properties of alumina microphotonic crystals with a diamond structure. *Journal of Materials Research*, **23(4)**, pp. 1036-1041, 2008.

[12] Kirihara, S., Uehara, Y., & Takinami Y., Fabrication of thermodynamic crystals by structural joining. *Proc. of the 34th international Conference Advanced Ceramics and Composites*, eds. T. Ohji & M. Singh, John Wiley & Sons: New Jersey, **31(8)**, pp. 169-174, 2010.

Section 3
Improvement in catalysis

Strong bond cleavage promoted by silyl group migration in a coordination sphere

H. Nakazawa
Department of Chemistry, Graduate School of Science, Osaka City University, Sumiyoshi-ku, Japan

Abstract

C-CN bond cleavage in organonitriles was attained in the photoreaction with $HSiEt_3$ in the presence of a catalytic amount of $(\eta^5\text{-}C_5H_5)Fe(CO)_2Me$. The reaction sequence was proposed, where silyl migration from Fe to N in the coordinating nitrile is a key step. DFT calculations proposed a silyl migration with small activation energy of 4.0 kcal/mol and subsequent C-CN bond cleavage on the Fe coordination sphere. N-CN bond cleavage in cyanamide (R_2NCN) was also attained by a transition metal catalyst. A catalytic cycle was proposed where a silyl migration from a transition metal to the nitrile nitrogen is involved. An N-silylated η^2-amidino complex was isolated, which was shown to be an intermediate in the catalytic pathway.
Keywords: C-C bond cleavage, C-N bond cleavage, transition metal catalysis, silyl migration.

1 Introduction

Investigations of selective bond cleavage as well as selective bond formation of chemical compounds is important from the viewpoint of atom efficiency, low environmental load, and sustainable chemistry. Weak bond cleavage is not so difficult, whereas strong bond cleavage with keeping weak bonds intact is quite difficult. Therefore, selective strong bond cleavage is a challenging topic in chemistry. One of the promising ways to overcome the difficulties is using a transition metal catalyst.

As the carbon-carbon bond is relatively unreactive, C-C bond cleavage is an area of considerable current interest [1]. In particular, C-C bond activation in

WIT Transactions on Ecology and the Environment, Vol 154, © 2011 WIT Press
www.witpress.com, ISSN 1743-3541 (on-line)
doi:10.2495/CHEM110121

acetonitrile is difficult because it has rather strong C-C bond energy (133 kcal mol^{-1}) compared with alkane C-C bond energy (*ca.* 83 kcal mol^{-1}).

Transition metal complexes have been used to attain C-C bond cleavage of nitrile. They mainly involve Group 10 transition metal triads [2–10]. In addition, one example for Mo [11], Co [12], U [13], and two examples of Cu [14, 15] have been reported. These reactions show stoichiometric C-C bond cleavage. In contrast, a few catalytic reactions involving Ni [16] and Pd [17] complexes have been reported. For these examples, it is proposed or clearly shown that a direct oxidative addition of a C-CN bond toward an electronically unsaturated transition metal fragment takes place to give an alkyl(aryl)-cyano complex (eq. (1)).

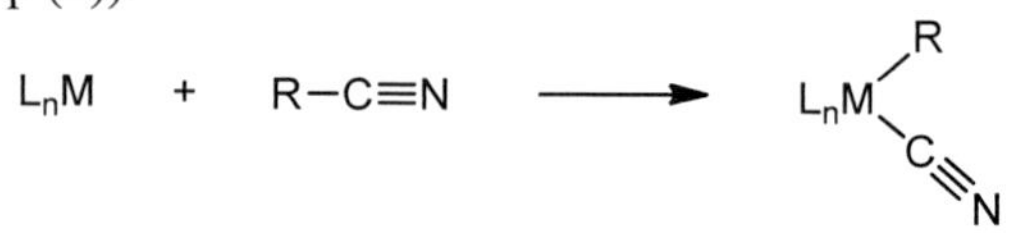

(1)

This article introduces a new reaction pathway for organonitrile C-C bond cleavage where silyl group migration from a transition metal to nitrile nitrogen in the coordination sphere is a key step [18–20]. This article also report silyl assisted N-CN bond cleavage of cyanamide [21].

2 Stoichiometric C-CN bond cleavage of acetonitrile by a silyl-iron complex

Photoreaction with a 400-W medium pressure mercury arc lamp (pyrex filtered) of a THF solution containing $CpFe(CO)_2(SiMe_3)$, MeCN, and 2 equiv of $P(NMeCH_2)_2(OMe)$ (L) generated $CpFe(CO)L(SiMe_3)$ (25% yield), $CpFe(CO)L(Me)$ (6% yield), $CpFeL_2(Me)$ (15% yield), $CpFeL_2(CN)$ (19% yield), and Me_3SiCN (66% yield) (eq. (2)). When CD_3CN was used in place of CH_3CN, the corresponding CD_3-iron complexes ($CpFe(CO)L(CD_3)$ and $CpFeL_2(CD_3)$) were obtained. These results obviously show that C-C bond in acetonitrile is cleaved in this reaction. In order to check whether the silyl group on the iron plays an important role to cleave the C-C bond in acetonitrile, $CpFe(CO)_2Me$, $CpFe(CO)_2(GeMe_3)$, and $CpFe(CO)_2(SnMe_3)$ were used in place of $CpFe(CO)_2(SiMe_3)$. No C-CN bond cleavage reaction took place, indicating that the silyl ligand on the iron is inevitable to cleave C-CN bond in acetonitrile [18].

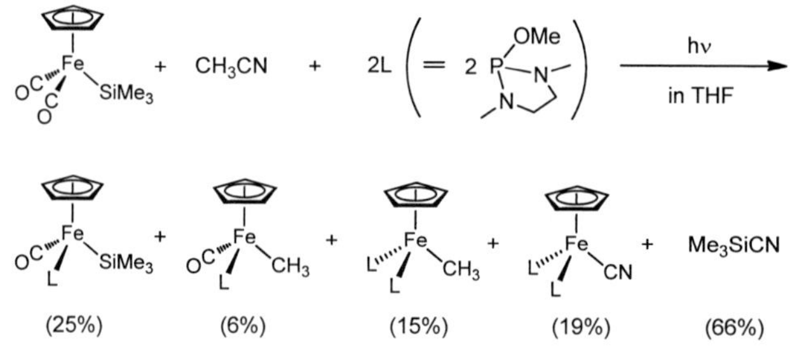

(2)

Similar silyl-assisted nitrile C-C bond cleavage was reported independently by Bergman, Brookhart and co-workers using Rh [22] and Ir [23] complexes. Later, Hashimoto et al. [24] and Tobiso et al. [25] and their co-workers reported this type of silyl-assisted nitrile C-C bond cleavage.

3 Theoretical approach

In order to elucidate the role of a silyl group in the iron complex toward C-C bond activation of acetonitrile, DFT calculations were performed. It is expected that at first the irradiation of the coordinatively saturated $CpFe(CO)_2(SiMe_3)$ will lead to the dissociation of one CO ligand to form 16e species $CpFe(CO)(SiMe_3)$ (**A**). The direct C-C bond oxidative addition of acetonitrile to the 16e species is less likely to happen, because this reaction requires a very high activation barrier of ca. 53 kcal/mol. Several DFT calculations could find a plausible reaction path which is shown in Figure 1 [18].

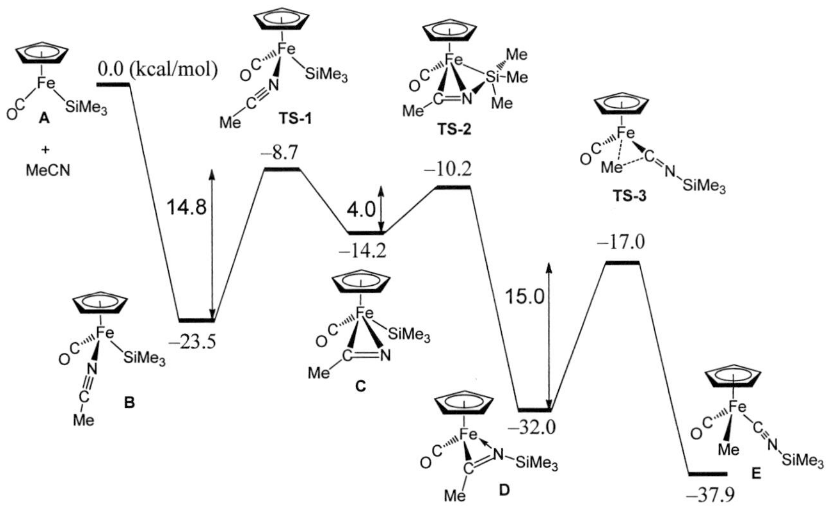

Figure 1: Energy profiles of the reaction of $CpFe(CO)(SiMe_3)$ with MeCN (kcal.mol).

The reaction of **A** with MeCN forms **B** with the end-on coordination of acetonitrile with releasing 23.5 kcal/mol. Then, **B** is converted into the CN π-coordinated species **C**. This process takes place through a transition state (**TS-1**) with the activation energy of 14.8 kcal/mol. In **C**, the migration of the silyl group from Fe to nitrile nitrogen is possible. The calculations showed that this requires only a small activation energy of 4.0 kcal/mol (**C** to **TS-2**). From **TS-2**, a somewhat stable imino complex with the coordination of a nitrogen lone pair to the Fe atom, **D**, is formed. If the methyl group of the imino ligand migrates to the iron atom, the methyl iron complex (**E**) could be formed. It has been found that a distortion in **D** that breaks the Fe-N bond leads to the transition state **TS-3**. This process requires an activation energy of 15.0 kcal/mol. The product of this step, **E**, is 5.9 kcal/mol more stable than **D**, which is presumably due to the presence of a strong Fe-C bond.

The reaction profile in Scheme 1 well explains the rupture of the acetonitrile C-C bond by the Fe(II) complex. The highest activation barrier of 15 kcal/mol is well within the reach of a feasible chemical reaction. Once **E** is formed, the formation of CpFe(CO)LMe can be easily achieved by a silylisocyanide/L exchange reaction. The trimethylsilyl isocyanide dissociated from **E** may isomerize to trimethylsilyl cyanide. Theoretical calculation suggests that the cyanide is more stable than the isocyanide. This is in accord with the experimental fact that trimethylsilyl cyanide is thermodynamically more favorable [26].

4 Catalytic C-CN bond cleavage of acetonitrile

In order to establish a catalytic cycle, photoreaction of a THF solution containing an equimolar amount of Et_3SiH and MeCN in the presence of a catalytic amount of $CpFe(CO)_2(SiMe_3)$ (2 mol%) was examined. The main Si-containing product was Et_3SiCN. The yield based on Et_3SiH used was 45% (TON = 23). The results show the C-C bond in MeCN is cleaved catalytically.

Table 1: Catalytic activities of various iron complexes.[a]

$$Et_3SiH + H_3C\text{-}CN \xrightarrow[\text{THF, } h\nu]{\text{catalyst}} Et_3SiCN + CH_4$$

Entry	Catalyst	TON[b]
1	$Cp(CO)_2Fe(SiMe_3)$	22.5
2	$Cp(CO)_2Fe(benzyl)$	21.3
3	$Cp(CO)_2FeMe$	36.0
4	$Cp(CO)_2FeCl$	0.6
5	$Cp(CO)_2FeI$	2.0
6	$(C_5MeH_4)(CO)_2FeMe$	32.5
7	$(C_5Me_4H)(CO)_2FeMe$	5.7
8	$(C_5Me_5)(CO)_2FeMe$	4.3
9	$(C_5H_4SiMe_3)(CO)_2FeMe$	26.4
10	$(C_5H_4SiMe_3)(CO)_2FeI$	2.0
11	$(C_5H_4I)(CO)_2FeMe$	12.6
12	$(C_5H_4\{P(O)(OMe)_2\})(CO)_2FeMe$	9.8

[a] Reactions were carried out at room temperature for 24 h under photoirradiation by using catallyst (0.52 mmol), Et_3SiH (26.00 mmol), and THF (5.00 mL) in acetonitrile (13.6 mL, 260.00 mmol). [b] Determined by GC with toluene as internal standard.

Next, the effect of a ligand at the Fe center on the catalytic activity was examined. For $CpFe(CO)_2R$ the methyl complex showed better activity than the silyl, benzyl, halo complexes. For $(C_5R_5)Fe(CO)_2Me$ introduction of substituents into the Cp ring reduced the catalytic activity. therefore, $CpFe(CO)_2Me$ showed the best catalytic activity among complexes in Table 1. The catalytic activities of the related methyl complexes of Mo and W, $CpM(CO)_3Me$ (M = Mo, W), were sluggish.

5 Catalytic C-CN bond cleavage of organonitriles other than acetonitrile

To see the scope and limitation of catalytic activity of $CpFe(CO)_2Me$ for C–CN bond cleavage, reactions with several organonitriles other than acetonitrile were examined. Table 2 summarizes the results. Propionitrile (EtCN) was converted into Et_3SiCN in 73% yield (TON: 18.2) (Table 2, entry 1). Isobutyronitrile (^{i}PrCN) and malononitrile ($NCCH_2CN$) were poorly converted (Table 2, entries 2, 3). However, succinonitrile ($NCCH_2CH_2CN$) was converted as much as propionitrile (65% yield, TON = 16.3; Table 2, entry 4).

Organonitriles in entries 5-10 were resistant to the C-CN bond cleavage in the reaction conditions (Table 2, entries 5-11). These results indicate that an electron-withdrawing, bulky, or coordination-feasible substituent on a carbon adjacent to the CN group is unfavorable for the C-CN bond cleavage.

Table 2: R-CN bond cleavage reactions of organonitriles.[a]

Entry	1	2	3[b]	4[b]	5	6	7	8	9
R		^{i}Pr	NC	NC			^{t}Bu	Cl_3C	Me_2N
TON	18.2	4.2	7.4	16.3	NR	NR	NR	0.3	NR

Entry	10	11	12[c]	13[c]	14[c]	15[c]	16[c]	17[c]
R	MeO	o-tolSO$_2$		CN	CN	NC	MeO	F_5
TON	trace	0.4	12.2	10.5	11.7	11.5	15.2	trace

[a] Catalyst (0.20 mmol), Et_3SiH (5.00 mmol), THF (0.40 mL), nitrile compound (50.00 mmol). [b] Nitrile (2.50 mmol), THF (1.30 mL). [c] Catalyst (0.20 mmol), Et_3SiH (5.00 mmol), THF (8.00 mL), nitrile (5.00 mmol). [d] Determined by GC with toluene as internal standard.

For aryl nitriles, the C-CN bonds were cleaved with TONs of about 10 (Table 2, entries 12-15). The TON for phthalonitrile is slightly lower (Table 2, entry 13) than that for benzonitrile, which may be due to the presence of an electron-withdrawing group in the *ortho* position. The TON for *p*-methoxybenzonitrile is slightly higher than for benxonitrile (Table 2, entry 16), which may be due to the electron-releasing OMe group. The C-CN bond was hardly cleaved for pentafluorobenzonitrile (table 2, entry 17), possibly due to strongly electronegative nature of F.

6 Consideration of catalytic cycle

A plausible catalytic cycle is shown in Figure 2 for the reaction of MeCN with Et_3SiH in the presence of $CpFe(CO)_2Me$ as a catalyst precursor. One CO ligand in the precursor is released by photolysis to give CpFe(CO)Me, which reacts with Et_3SiH to give $CpFe(CO)(Me)(H)(SiEt_3)$. This process seems plausible because oxidative addition of an Si-H bond to a 16e-species of Fe was reported to give, for example, $CpFe(CO)(H)(SiEt_3)_2$ [27]. The subsequent reductive elimination of CH_4 yields $CpFe(CO)(SiEt_3)$. As $SiEt_3Me$ was not observed at all in this system, reductive elimination of CH_4 from $CpFe(CO)Me(H)(SiEt_3)$ seems to precede that of Et_3SiMe. The 16e species thus formed reacts with MeCN to give ultimately $CpFe(CO)(Me)(\eta^1\text{-}CNSiEt_3)$. The reaction sequences have been demonstrated theoretically (Figure 1) [18]. Dissociation of Et_3SiNC generates CpFe(CO)Me to complete the catalytic cycle. The released Et_3SiNC isomerizes to thermodynamically stable Et_3SiCN.

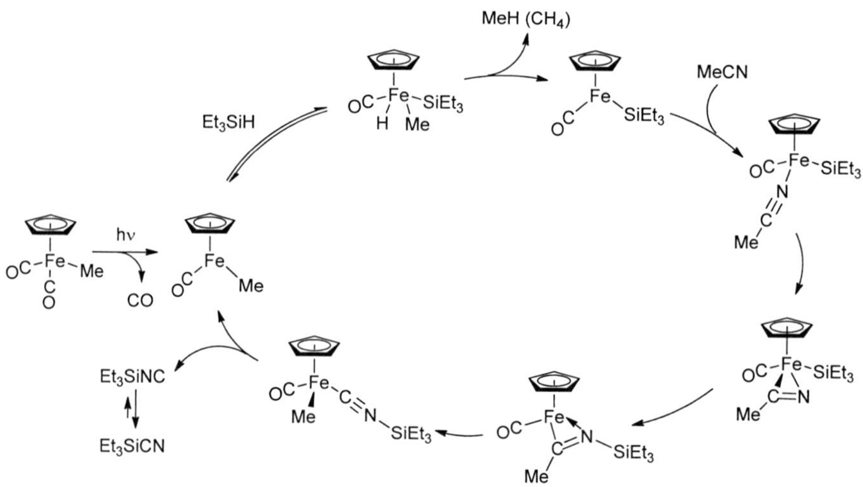

Figure 2: Proposed catalytic cycle.

According to the catalytic cycle shown in Figure 2, $CpFe(CO)(SiEt_3)$ can react with both MeCN and Et_3SiH. The former reaction seems dominant, but the latter is not negligible. If the reaction of $CpFe(CO)(SiEt_3)$ with Et_3SiH is suppressed, the catalytic cycle is expected to work more effectively. Thus the reaction on changing the molar ratio of Et_3SiH and MeCN was examined. Reaction of Et_3SiH with a 10-fold molar excess of MeCN in the presence of 0.83 mol% $CpFe(CO)_2Me$ under photolysis for 24 h at 50°C produced 99% yield of Et_3SiCN base on Et_3SiH (TON: 118). The TON increased with the photoirradiation time (48 h: 156, 96 h: 197). The highest TON (251) was obtained when Et_3SiH and a 10-fold molar excess of MeCN were photolyzed for

1 week at 50°C in the presence of 0.2 mol% of $CpFe(CO)_2Me$. Compared with Pd [17] and Rh [22] systems, this catalytic system is very effective.

7 N-CN bond cleavage of cyanamides

New reaction pattern of C-CN bond cleavage of organonitriles promoted by a silyl-iron complex has been described above. The essence of the reaction mechanism is silyl migration from Fe to the nitrogen in nitrile which is η^2-coordinated through C≡N π-bond. This migration causes C-CN bond cleavage. The replacement of the R group in RCN (organonitrile) by an NR_2 group yields cyanamide described as R_2NCN. Therefore, there is a possibility of N-CN bond cleavage of cyanamide in the reaction with a transition metal complex bearing a silyl ligand.

The R_2N-CN bond is known to be strong and not broken readily. For example, the N-CN bond length in Me(p-C_6H_4Cl)N-CN is reported to be 1.331 Å, which lies just between that of a normal N-C single bond (1.47 Å) and that of an N=C double bond (1.27 Å) [28]. The von Braun reaction is the only reaction known to date to cleave R_2N-CN bond. However, it requires harsh reaction conditions (strong acid or base conditions). Reactions of cyanamides with a transition metal complex bearing a silyl ligand were examined both in hopes of N-CN bond cleavage and with worrying about coordination of cyanamide to the 16e Fe species, $Cp(CO)Fe(SiEt_3)$, through a lone pair of electrons on the amino nitrogen causing disappearance of the activity of the iron complex toward R_2N-CN bond cleavage.

Table 3: Photoreaction of cyanamides with $CpFe(CO)_2(SiEt_3)$.

$$R_2N\text{-}CN + CpFe(CO)_2(SiEt_3) \xrightarrow[\text{toluene}]{h\nu,\ \text{at r.t.}} Et_3SiCN$$

Entry	Substrates	Time/h	Yield (%) [a]
1	Me_2NCN	12	51
2	^{n}Hex_2NCN	12	30
3	N-cyanopiperidine	12	41
4	N-cyanomorpholine	12	32
5	N-cyanopyrrolidine	12	26
6	H_2NCN	12	20[b]
7	$Me_2N(BH_3)CN$	24	14
8	$Me_2N(BF_3)CN$	24	0
9	$C_5H_{10}N(BH_3)CN$	24	18

[a] Yield of Et_3SiCN obtained by GC. [b] In 1,2-dichloroethane

A solution of an equimolar amount of dimethylcyanamide and $CpFe(CO)_2(SiEt_3)$ (**1**) in toluene was photoirradiated at room temperature for 12 h. The ^{1}H NMR spectrum and the GC analysis of the reaction mixture showed formation of Et_3SiCN. The yield was 51% (Table 3, entry 1), showing that the

Me_2N-CN bond cleavage could be attained at room temperature using a silyl-iron complex.

Results of reactions with other cyanamides are presented in Table 3. Although the yields of Et_3SiCN are less than 50%, these N-CN bonds are cleaved (entries 2-6). The reaction of H_2NCN is noteworthy (entry 6). The H_2N-CN bond has a double bond character because H_2N-CN (cyanamide) is a tautomer of HN=C=NH (carbodiimide). Therefore H_2N-CN bond is stronger than other R_2N-CN. The first H_2N-CN bond cleavage is attainable in these reaction conditions, although the efficiency remains insufficient.

Derivation of cyanamide into the borane adduct at the amino nitrogen, $R_2N(BX_3)CN$ (X = H, F) [29], might engender more effective R_2N-CN bond cleavage because of masking of the lone pair electrons on the amino nitrogen. The results (entries 7-9) showed that the introduction of borane into cyanamide did not facilitate R_2N-CN bond cleavage; instead, it reduced the activity, presumably because of steric hindrance.

Reaction sequences resembling those in Figure 2 are expected for the reaction of **1** with cyanamide. Isolation of the N-silylated η^2-amidino iron intermediate was attempted, but the reaction of **1** with Me_2NCN was unsuccessful. However, reactions with Me_2NCN of $(C_5R_5)Fe(CO)(py)(SiR'R''_2)$ (py = pyridine), considered as a synthon of a 16e complex $(C_5R_5)Fe(CO)(SiR'R''_2)$, led to isolation of N-silylated η^2-amidino iron complexes (eq. (3)). Heating a solution containing **2** and Me_2NCN in benzene at 50 ºC for 10 h yielded **3** quantitatively according to the NMR measurements, but the isolation as a solid was failed. In contrast, a reaction of **4** with Me_2NCN yielded **5**, which could be isolated as dark-red powders in 85% yield. The unprecedented η^2-amidino complex was confirmed using X-ray analysis (Figure 3). The iron takes a distorted three-legged piano-stool structure with an η^2-amidino fragment. The bond distance of N1-C2 (1.327 Å) is shorter than that of a typical N-C single bond (*e.g.*, C3-N1 = 1.455 Å, C4-N1 = 1.458 Å), and is rather similar to that of an N=C double bond (*e.g.*, N2-C2 = 1.303 Å). The sum of angles around N1 is 359.9º. These structural characters are consistent with sp^2 hybridization of N1. The C3-N1-C2-N2 fragment is nearly planar with a torsion angle of 2.6(4)º. Both ^1H and ^{13}C NMR spectra show that the structure in a solid state is maintained in solution. Two NCH_3 resonances were observed at room temperature in ^1H and ^{13}C NMR, reflecting that the C2-N1 bond does not rotate freely at room temperature.

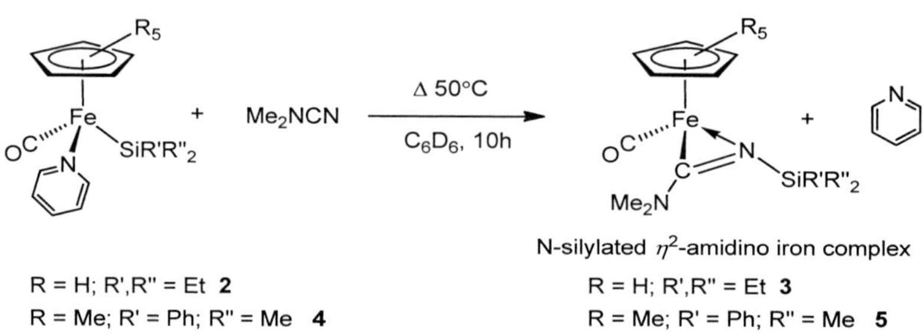

(3)

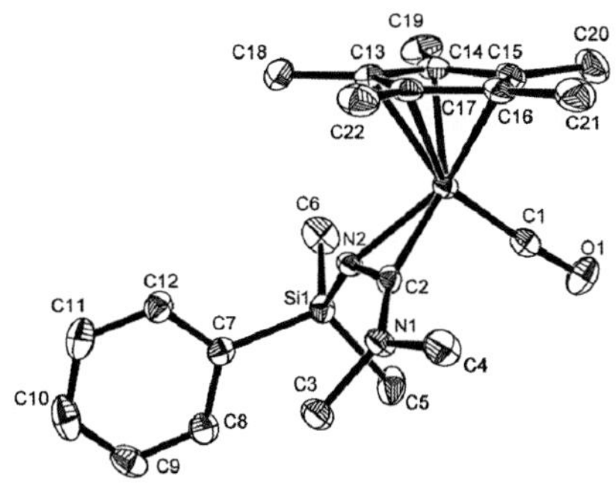

Figure 3: ORTEP drawing of **5**. Selected bond distances (Å) and angles (deg): N1-C2 = 1.327 (3); N2-C2 = 1.303(2); C2-N1-C3 = 123.4(2); C2-N1-C4 = 119.5(2); C3-N1-C4 = 117.0(2).

Complexes **3** and **5** were subjected to a thermal reaction. Although **5** produced a small amount of $PhMe_2SiCN$ on heating in toluene at 110°C for 24 h, **3** gave Et_3SiCN in 62% yield on heating in benzene at 70°C for 24 h. The results show that an N-silylated η^2-amidino complex is an intermediate in the N-CN bond cleavage of cyanamide.

Table 4: Catalytic cleavage under photolysis or heating.

$$\mathrm{Me_2NCN} + \mathrm{Et_3SiH} \xrightarrow[\text{toluene}]{\text{cat. } \mathbf{6} \text{ or } \mathbf{7}^a} \mathrm{Et_3SiCN}$$

Entry	Cat.	[M]:[N]:[Si][b]	Condition	Temp/°C	Time/h	TON[c]
1	**6**	1 : 10 : 10	hν	25	24	0.4
2	**7**	1 : 10 : 10	hν	25	24	1.4
3	**6**	1 : 1 : 1	Δ	80	12	0
4	**7**	1 : 1 : 1	Δ	100	12	0.52
5[d]	**7**	1:10:1000	Δ	100	48	7.9
6[d]	**7**	1:1000:5000	Δ	100	120	32.3

[a] No Me_2NCN bond cleavage took place in the absence of **6** or **7**. [b] Molar ratio of a transition metal complex, Me_2NCN, and Et_3SiH. [c] Calculated from the isolated Et_3SiCN. The values are based on the concentration of a transition metal complex. [d] In free solvent.

Catalytic N-CN bond cleavage of cyanamides has been attempted. Table 4 shows the results of reactions of Me_2NCN and Et_3SiH in the presence of a catalytic amount of $CpFe(CO)_2Me$ or $CpMo(CO)_3Me$. Entry 1 shows that the Fe complex does not work as a catalyst, whereas the Mo complex does under photo-irradiation conditions (Table 4, entry 2). The Mo complex shows catalytic activity even under thermal conditions (Table 4, entries 4-6). These results show the first example of catalytic N-CN bond cleavage of cyanamides by a transition metal complex.

8 Conclusions

It was found that a 16e species bearing a transition metal-silyl bond, such as $CpFe(CO)(SiR_3)$ and $CpMo(CO)_2(SiR_3)$, serves as a real catalyst for C-CN and N-CN bond cleavage in organonitriles and cyanamides, respectively. Silyl migration from a transition metal to the nitrogen of the nitrile group in RCN and R_2NCN coordinating to the transition metal in η^2-fashion leads to the strong C-CN and N-CN bond cleavage.

References

[1] a) Crabtree, R. H., The organometallic chemistry of alkanes. *Chem. Rev.* **85**, pp. 245-269, 1985. b) Rybtchinski, B. & Milstein, D., Metal insertion into C-C bonds in solution. *Angew. Chem., Int. Ed.*, **38**, pp. 870-883, 1999. c) Jun, C.-H., Transition metal-catalyzed carbon-carbon bond activation. *Chem. Soc. Rev.*, **33**, pp. 610-618, 2004.

[2] Burmeister, J. L. & Edwards, L. M., Carbon-carbon bond cleavage via oxidative addition: Reaction of tetrakis(triphenylphosphine)platinum(0) with 1,1,1-tricyanoethane. *J. Chem. Soc. A.*, pp. 1663–1666, 1971.

[3] Gerlach, D. H., Kane, A. R., Parshall, G. W., Jesson, J. P. & Muetterties, E. L., Reactivity of trialkylphosphine complexes of platinum(0). *J. Am. Chem. Soc.*, **93**, pp. 3543–3544, 1971.

[4] Cassar, L., A new nickel-catalyzed synthesis of aromatic nitriles. *J. Organomet. Chem.*, **54**, pp. C57–C58, 1973.

[5] Parshall, G. W., σ-Aryl compounds of nickel, palladium, and platinum. Synthesis and bonding studies. *J. Am. Chem. Soc.*, **96**, pp. 2360–2366, 1974.

[6] Favero, G., Morvillo, A. & Turco, A., Oxidative addition of alkanenitriles to nickel(0) complexes via π-intermediates. *J. Organomet. Chem.*, **241**, pp. 251–257, 1983.

[7] Bianchini, C., Masi, D., Meli, A. & Sabat, M., Reactions of ethyl cyanoformate with $(np_3)Ni$, $(np_3)CoH$ [$np_3 = N(CH_2CH_2PPh_2)_3$], and (triphos)Ni [triphos = $MeC(CH_2PPh_2)_3$]. Crystal structures of the ethoxycarbonyl complexes $[(np_3)Ni(CO_2Et)]BPh_4$ and $(triphos)Ni(CN)(CO_2Et)$. *Organometallics*, **5**, pp. 1670–1675, 1986.

[8] Abla, M. & Yamamoto, T., Oxidative addition of C-CN bond to nickel(0) complex: Synthesis and crystal structures of $Ni(CN)(o\text{-}C_6H_4CN)(bpy)$ and $Ni(CN)(p\text{-}C_6H_4CN)(bpy)$. *J. Organomet. Chem.*, **532**, pp. 267–270, 1997.

[9] a) Brunkan, N. M., Brestensky, D. M. & Jones, W. D., Kinetics, thermodynamics, and effect of BPh_3 on competitive C-C and C-H bond activation reactions in the interconversion of allyl cyanide by [No(dippe)]. *J. Am. Chem. Soc.*, **126**, pp. 3627–3641, 2004. b) Garcia, J. J., Arevalo, A., Brunkan, N. M. & Jones, W. D., Cleavage of carbon-carbon bonds in alkyl cyanides using nickel(0). *Organometallics*, **23**, pp. 3997–4002, 2004.

[10] Liu, Q.-X., Xu, F.-B., Li, Q.-S., Song, H.-B. & Zhang, Z.-Z., Formation of the fluorescent complexes $[(carbene)_2M^{II}(CN)_2]$ (M = Ni, Pd, Pt) by C-C bond cleavage of CH_3CN. *Organometallics*, **23**, pp. 610–614, 2004.

[11] Churchill, D., Shin, J. H., Hascall, T., Hahn, J. M., Bridgewater, B. M. & Parkin, G., The *ansa* effect in permethylmolybdenocene chemistry: A $[Me_2Si]$ *ansa* bridge promotes intermolecular C-H and C-C bond activation. *Organometallics*, **18**, pp. 2403–2406, 1999.

[12] Ozawa, F., Iri, K. & Yamamoto, A., C-CN bond cleavage under mild conditions promoted by electron-rich cobalt complexes. *Chem. Lett.*, pp. 1707–1710, 1982.

[13] Adam, R., Villiers, C., Ephritikhine, M., Lance, M., Nierlich, M. & Vigner, J., Synthesis, structure and oxidative addition reactions of triscyclopentadienyluranium(III) nitrile complexes. *J. Organomet. Chem.*, **445**, pp. 99–106, 1993.

[14] Marlin, D. S., Olmstead, M. M. & Mascharak, P. K., Heterolytic cleavage of the C-C bond of acetonitrile with simple monomeric Cu^{II} complexes: Melding old copper chemistry with new reactivity. *Angew. Chem. Int. Ed.*, **40**, pp. 4752–4754, 2001.

[15] Lu, T., Zhuang, X., Li, Y. & Chen, S., C-C bond cleavage of acetonitrile by a dinuclear copper(II) cryptate. *J. Am. Chem. Soc.*, **126**, pp. 4760–4761, 2004.

[16] a) Penney, J. M. & Miller, J. A., Alkynylation of benzonitriles via nickel catalyzed C-C bond activation. *Tetrahedron Lett.*, **45**, pp. 4989–4992, 2004. b) Chaumonnot, A., Lamy, F., Sabo-Etienne, S., Donnadieu, B., Chaudret, B., Barthelat, J.-C. & Galland, J.-C., Catalytic isomerization of cyanoolefins involved in the adiponitrile process. C-CN bond cleavage and structure of the nickel π-allyl cyanide complex $Ni(\eta^3\text{-}1\text{-}Me\text{-}C_3H_4)(CN)(dppb)$. *Organometallics*, **23**, pp. 3363–3365, 2004. c) Wilting, J., Müller, C., Hewat, A. C., Ellis, D. D., Tooke, D. M., Spek, A. L. & Vogt, D., Nickel-catalyzed isomerization of 2-methyl-3-butenenitrile. *Organometallics*, **24**, pp. 13–15, 2005. d) Nakao, Y., Yukawa, T., Hirata, Y., Oda, S., Satoh, J. & Hiyama, T., Allylcyanation of alkynes: Regio- and stereoselective access to functionalized di- or trisubstituted acrylonitriles. *J. Am. Chem. Soc.*, **128**, pp. 7116–7117, 2006.

[17] a) Murahashi, S., Naota, T. & Nakajima, N., Palladium-catalyzed decarbonylation of acyl cyanides. *J. Org. Chem.*, **51**, pp. 898–901, 1986. b) Luo, F.-H., Chu, C.-I. & Cheng, C.-H., Nitrile-group transfer from solvents to aryl halides. Novel carbon-carbon bond formation and cleavage mediated by palladium and zinc species. *Organometallics*, **17**, pp. 1025–1030, 1998. c) Nishihara, Y., Inoue, Y., Itazaki, M. & Takagi, K., Palladium-catalyzed cyanoesterification of norbornenes with cyanoformates via the NC-Pd-COOR (R = Me and Et) intermediate. *Org. Lett.*, **7**, pp. 2639–2641, 2005.

[18] Nakazawa, H., Kawasaki, T., Miyoshi, K., Suresh, C. H. & Koga, N., C-C bond cleavage of acetonitrile by a carbonyl iron complex with a silyl ligand. *Organometallics*, **23**, pp. 117–126, 2004.

[19] Nakazawa, H., Kamata, K. & Itazaki, M., Catalytic C-C bond cleavage and C-Si bond formation in reaction of RCN with Et_3SiH promoted by an iron complex. *Chem. Commun.*, pp. 4004–4006, 2005.

[20] Nakazawa, H., Itazaki, M., Kamata, K. & Ueda, K., Iron-complex-catalyzed C-C bond cleavage of organonitriles: Catalytic metathesis reaction between H-Si and R-CN bonds to afford R-H and Si-CN bonds. *Chem. Asian. J.*, **2**, pp. 882-888, 2007.

[21] Fukumoto, K., Oya, T., Itazaki, M. & Nakazawa, H., N-CN bond cleavage of cyanamides by a transition-metal complex. *J. Am. Chem. Soc.*, **131**, pp. 38-39, 2009.

[22] Taw, F. L., Mueller, A. H., Bergman, R. G. & Brookhart, M., A mechanistic investigation of the carbon-carbon bond cleavage of aryl and alkyl cyanides using a cationic Rh(III) silyl complex. *J. Am. Chem. Soc.*, **125**, pp. 9808–9813, 2003.

[23] Klei, S. R., Tilley, T. D. & Bergman, R. G., Stoichiometric and catalytic behavior of cationic silyl and silylene complexes. *Organometallics*, **21**, pp. 4648–4661, 2002.

[24] Hashimoto, H., Matsuda, A. & Tobita, H., Reactions of a silyl(silylene)iron complex with nitriles: Carbon-carbon bond cleavage of nitriles by the transiently generated disilanyliron(II) intermediate. *Organometallics*, **25**, pp. 472–476, 2006.

[25] Tobisu, M., Kita, Y. & Chatani, N., Rh(I)-catalyzed silylation of aryl and alkenyl cyanides involving the cleavage of C-C and Si-Si bonds. *J. Am. Chem. Soc.*, **128**, pp. 8152–8153, 2006.

[26] Seckar, J. A. & Thayer, J. S., Normal-iso rearrangement in cyanotrialkylsilanes. *Inorg. Chem.*, **15**, pp. 501-504, 1976.

[27] Akita, M., Oku, T., Tanaka, M. & Moro-oka, Y., Oxymethylation of alkyliron complex $CpFe(CO)_2R$ with hydrostannane and -silane, leading to RCH_2OH derivatives: related reactions of CpFe(CO)(L)R and CpFe(CO)(L)C(O)R type organoiron complexes and the molecular structure of trans-$CpFe(H)(CO)(SnPh_3)_2$. *Organometallics*, **10**, pp. 3080–3089, 1991.

[28] Cunningham, I. D., Light, M. E. & Hursthouse, M. B., N-(4-chlorophenyl)-N-methylcyanamide. *Acat Cryst. Sect. C, Cryst. Struct. Commun.*, **55**, pp. 1833-1835, 1999.

[29] Henneike, H. F. & Drago, R. S., Comparison of the donor properties of dimethylcyanamide and acetonitriles. *Inorg. Chem.*, **7**, pp. 1908-1915, 1968.

Bridging the gap between cellulose chemistry and heterogeneous catalysis

S. Van de Vyver[1], J. Geboers[1], L. Peng[1], F. de Clippel[1], M. Dusselier[1], T. Vosch[2], L. Zhang[3], G. Van Tendeloo[3], C. J. Gommes[4,5], B. Goderis[5], P. A. Jacobs[1] & B. F. Sels[1]
[1]*Center for Surface Chemistry and Catalysis, Katholieke Universiteit Leuven, Belgium*
[2]*Department of Chemistry, Katholieke Universiteit Leuven, Belgium*
[3]*Center for Electron Microscopy for Materials Science, University of Antwerp, Belgium*
[4]*Department of Chemical Engineering, University of Liège, Belgium*
[5]*Department of Molecular and Nanomaterials, Katholieke Universiteit Leuven, Belgium*

Abstract

Although cellulosic biomass offers a promising alternative as an abundant renewable resource in the production of biofuels and platform chemicals, so far only a few studies have reported its aqueous-phase conversion into glucose or sugar alcohols using solid chemocatalysts. The principal reason is that these polymeric biomolecules with semi-crystalline structure cannot penetrate the pores of conventional heterogeneous chemocatalysts. New advances in the conversion of cellulose thus require the design of efficient multifunctional catalytic systems with sterically accessible acid and metal sites.
Keywords: biomass, carbon nanofibers, cellulose, glucose, heterogeneous catalysis, hydrogenation, hydrolysis, nanocomposites, nickel, sugar alcohols.

1 Introduction

Environmental concerns about the depletion of fossil fuel reserves, the impact of anthropogenic CO_2 emissions and increasing energy demands have encouraged the exploration of new catalytic methods for converting cellulose into valuable

www.witpress.com, ISSN 1743-3541 (on-line)
doi:10.2495/CHEM110131

platform chemicals and biofuels [1, 2]. The development of these sustainable catalytic transformations could potentially provide a long-term solution to the industrial dependence on fossil carbon, requiring in 2025 production of up to 30% of raw materials for the chemical industry from renewable resources [3, 4]. With an abundance of approximately 720 billion tonnes, *i.e.* 40% of the annual net yield of photosynthesis, cellulose is the world's largest organic raw material resource [5]. The blueprints of the "new" sustainable cellulose chemistry are based on some key elements, namely controlled depolymerization and catalytic cascade reactions (*e.g.*, hydrogenation, hydrogenolysis), which, when put together, yield a pool of molecules that can be used for synthesis of industrial intermediates and fine chemicals. Van de Vyver *et al.* [1] recently reviewed new advances in the conversion of cellulose, requiring the design of multifunctional catalysts with sterically accessible acid and metal sites. Our work provides an example for how two solid catalytic systems can be used for: (i) the hydrolysis of cellulose to glucose and (ii) the hydrolytic hydrogenation of cellulose to sugar alcohols such as sorbitol and mannitol. Before the two catalytic strategies are discussed, characterization of the substrate as well as a mechanical pretreatment method to decrease cellulose crystallinity will be addressed.

1.1 Pretreatment and characterization of cellulose

The potential of mechanical disruption of cellulose by breaking hydrogen bonds, taking into account integration with hydrolytic catalytic processes, is now being thoroughly studied by our research group [6–9]. In the present study, cellulose (Avicel PH-101, microcrystalline) was characterized (i) as received and (ii) after 24 h ball-milling pretreatment. Ball-milling with 25 g of cellulose powder was carried out using zirconia balls (mass 7.5 g; diameter 1.8 cm). The micrographs in Figure 1, taken with a high resolution scanning electron microscope (SEM; Philips XL-30 FEG), provide clear information about cellulose particle size distributions before and after ball-milling.

X-ray diffraction (XRD) patterns of the Ni/CNF catalyst were recorded at room temperature with a STOE STADI P Combi diffractometer. The diffracted intensity of CuKα radiation (wavelength of 0.154 nm) was measured in a 2θ

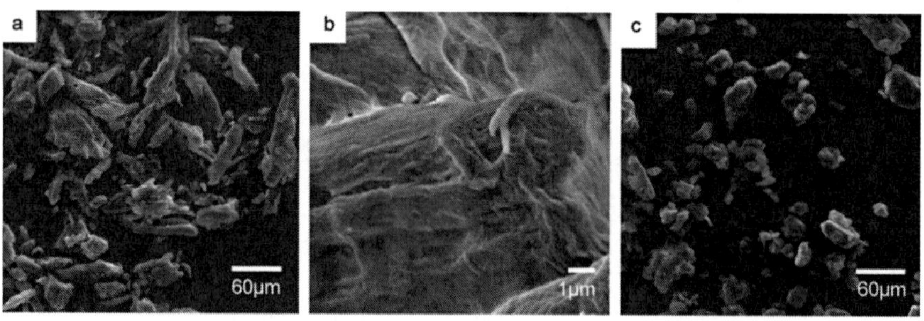

Figure 1: SEM images of Avicel cellulose, before (average particle size 50 µm; a, b) and after ball-milling (average particle size 20 µm; c).

range between 5 and 75°. Figure 2 displays XRD patterns taken of cellulose samples unmilled and ball-milled for 24 h. In the untreated cellulose, the major peak at $2\theta = 22.5°$ can be assigned to the crystalline plane 002. A comparison of the patterns clearly reveals a decrease in crystallinity of the cellulose feedstock after 24 h ball-milling pretreatment.

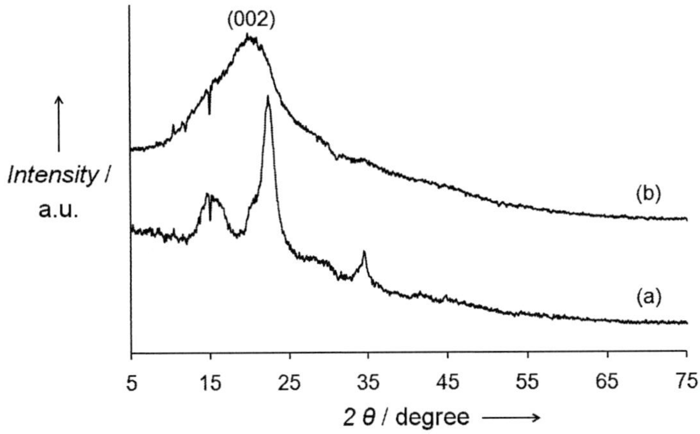

Figure 2: XRD patterns of cellulose before (a) and after ball-milling (b).

The ^{13}C CP MAS NMR spectra of cellulose with and without ball-milling were recorded on a Bruker Avance DSX400 spectrometer (9.4 T). The decreasing peak ratios of $C4_{(86\text{-}92\ ppm)}/C4_{(79\text{-}86\ ppm)}$ and $C6_{(63\text{-}67\ ppm)}/C6_{(56\text{-}63\ ppm)}$ in Figure 3 are in good agreement with the observations by Zhao *et al.* [10]. This, in interplay with the above XRD data, strongly suggests that more disordered cellulose is produced after ball-milling. Line widths are larger in the amorphous sample which is due to the large distribution of molecular orientations.

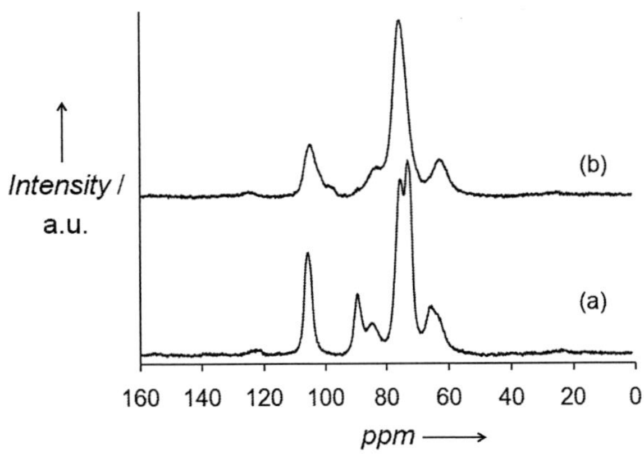

Figure 3: ^{13}C CP MAS NMR of cellulose before (a) and after ball-milling (b).

IR spectra were recorded under vacuum from KBr pellets on a Bruker IFS 66v/S instrument. The spectra in Figure 4 also show the changes in the cellulose structure after ball-milling. The less pronounced band at 1430 cm^{-1} is another strong indication of a less ordered cellulose sample, since it is assigned to the CH_2 scissoring motion in the cellulose I crystal. The ball-milling allows the regular arrangement of the CH_2OH group on C_6 to relax into a more random one, resulting in a broader band at 1430 cm^{-1}.

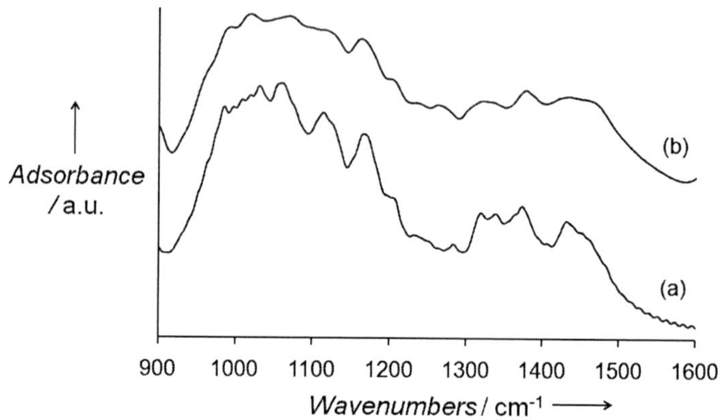

Figure 4: IR spectra of cellulose before (a) and after ball-milling (b).

1.2 Hydrolysis of cellulose to glucose

An excellent review on cellulose hydrolysis as entry point into biorefinery schemes has been published by Rinaldi and Schüth [11]. In the search for an alternative solid catalyst for cellulose hydrolysis, Van de Vyver *et al.* [6] recently reported the use of a new class of sulfonated silica/carbon nanocomposites, capable of achieving high glucose yields compared to reference ion-exchange resins (Amberlyst® 15, Nafion® NR50) and commercial zeolites.

1.3 Catalyst synthesis

The hybrid materials were synthesized by the evaporation-induced triconstituent co-assembly method, wherein sucrose is used as an alternative carbon precursor, prehydrolyzed tetraethyl orthosilicate (TEOS) as silica precursor, and Pluronic F127 triblock copolymer ($EO_{106}PO_{70}EO_{106}$, M_w = 12600) F127 as structure-directing amphiphilic surfactant (Figure 5). In a typical preparation procedure, 6.4 g block copolymer F127 was dissolved in 32 g ethanol with 0.3 g concentrated HCl (37%) and the mixture was treated in an ultrasonic bath for 1 h to afford a clear solution. Next, 8.32 g TEOS and 10 g sucrose solution with different concentration were added in sequence. It took 20 h at 313 K to evaporate ethanol and 24 h at 433K to thermopolymerize. Carbonization was done at 673K or 823K for 15 h under N_2 flow.

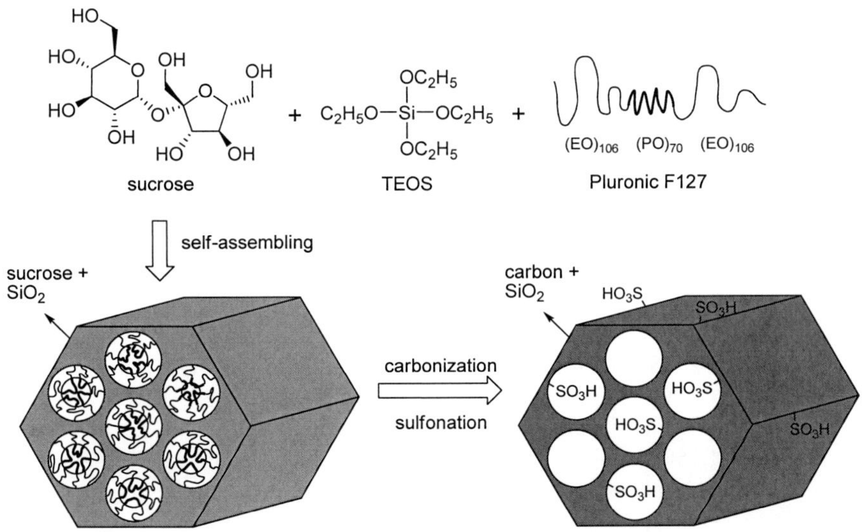

Figure 5: Schematic illustration of the proposed mechanism for formation of sulfonated silica/carbon nanocomposites (TEOS = tetraethyl-orthosilicate).

Three samples with different carbon content of 66wt%, 50wt% and 33wt% were synthesized by changing the concentration of the sucrose solution added. These materials were treated with concentrated sulfuric acid (1 g of solid/20 ml of H_2SO_4) at 423 K for 15 h in a Teflon-lined autoclave. The resulting materials were washed with hot distilled water until no sulfate ions were detected in the wash water. The as-prepared sulfonic acid functionalized silica/carbon nanocomposites are further denoted as SimCn-T-SO_3H (m, n are the weight percentages of silica and carbon, respectively, and T is the carbonization temperature expressed in Kelvin).

1.4 Catalyst characterization

The Small-Angle X-ray Scattering (SAXS) patterns of the as-synthesized and sulfonated samples were measured at the Dutch-Belgian beamline (BM26) of the European Synchrotron Radiation Facility and a custom software was used to reduce the measured 2D data into 1D scattering patterns [12]. Average distances between neighbouring pores were calculated as $a = 4\pi/(\sqrt{3}q^*)$, where q^* is the scattering vector of the first peak. The SAXS patterns of all samples in Figure 6 exhibit a clear maximum at scattering vector q^*, and dim secondary maxima are observed for some samples at larger q values. Although the exact positions of the latter cannot be determined accurately, they overlap nicely those expected for cylindrical pores being stacked according to a hexagonal (*p6mm*) symmetry, namely q^* and $2q^*$.

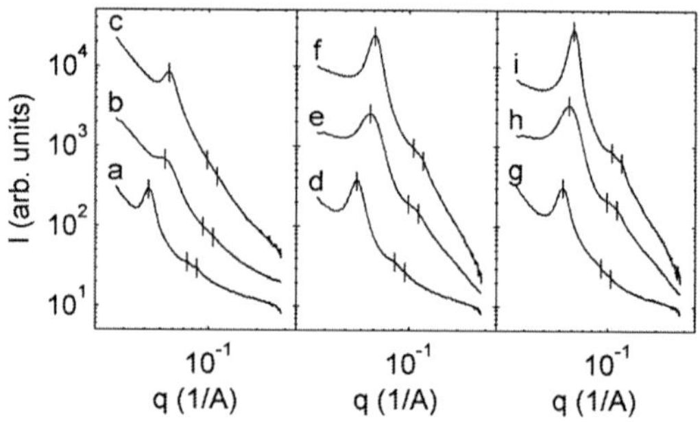

Figure 6: SAXS patterns of samples (a) Si33C66; (b) Si33C66-673-SO_3H; (c)Si33C66-823-SO_3H; (d) Si50C50; (e) Si50C50-673-SO_3H; (f)Si50C50-823-SO_3H; (g) Si66C33; (h) Si66C33-673-SO_3H; (i)Si66C33-823-SO_3H. The vertical lines indicate positions of the (1,0), (1,1) and (2,0) scattering peaks of a system having hexagonal symmetry [6].

The carbon content in the sulfonated silica/carbon nanocomposites was measured by thermogravimetric analysis (TGA) in O_2, as shown in Figure 7. TGA was performed on a TGA Q500 (TA Instruments, Brussels, Belgium). The samples were heated from room temperature to 1073 K at 5 K min^{-1}. For the sulfonic acid group-functionalized carbon-silica composite, the weight loss below 373 K and in range of 373–873 K in O_2 can be attributed to the absorbed water and carbon compound, respectively, while the weight residue belongs to the silica.

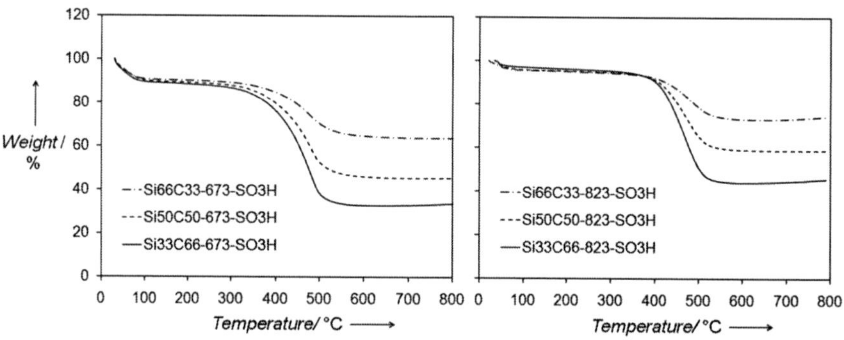

Figure 7: TGA curves of samples Si33C66-673-SO_3H, Si50C50-673-SO_3H, Si66C33-673-SO_3H, Si33C66-823-SO_3H, Si50C50-823-SO_3H and Si66C33-823-SO_3H.

1.5 Results and discussion

According to reference [6], Table 1 shows the acid density, glucose formation rate and turnover frequency (TOF) at acid sites for SimCn-T-SO_3H. The expectations based on titration results are in general agreement with the catalytic observations in that glucose formation is found to occur faster on nanocomposites with higher acid density.

More interestingly, the TOF data point to the prominent role that the hybrid-structured surface character plays in the hydrolytic activity of the Brønsted acid sites. Indeed, higher TOFs were determined for samples with increased Si/C mass ratio, and hence higher hydrophilicity. In previous studies it has been found that efficient conversion of cellulose requires a good interaction between the solid acid catalyst and β-1,4 glucan [1]. In the present case, the correlation between catalyst composition and TOF data strongly suggests that the hydrophilic silica groups facilitate substrate adsorption.

Table 1: Catalytic conversion of cellulose over sulfonated silica /carbonnanocomposites. Reaction conditions: cellulose pretreated by ball-milling 0.05 g, catalyst 0.05 g, water 5 mL, reaction time 24 h, temperature 423 K [6].

Catalyst	Acid density (mmol g^{-1})	TOF (h^{-1})	Conv. (%)	Glucose yield (%)
Si33C66-673-SO_3H	0.57	0.15	42.5	38.3
Si50C50-673-SO_3H	0.40	0.15	33.0	27.0
Si66C33-673-SO_3H	0.31	0.19	31.3	26.9
Si33C66-823-SO_3H	0.37	0.30	60.7	50.4
Si50C50-823-SO_3H	0.25	0.37	50.5	42.1
Si66C33-823-SO_3H	0.15	0.41	35.9	27.6

2 Hydrolytic hydrogenation of cellulose to sugar alcohols

Another elegant strategy to valorise cellulose is inspired by a previous patent application by Jacobs *et al.* [13] on catalytic starch conversion. Such one-pot catalytic approach relies on a proton-catalyzed hydrolysis of the glycoside bonds, followed by fast metal-catalyzed hydrogenation of glucose into sugar alcohols such as sorbitol. Most recently, Geboers *et al.* [8] developed a novel catalytic system for the hydrolytic hydrogenation of cellulose, consisting of trace amounts of mineral acids and a bifunctional Ru/H-USY or Ru/H-MOR catalyst. The reported system exhibits exceptional activity and product selectivity, resulting in fast cellulose conversion with very high hexitol yields (>90%). Geboers *et al.* [9] also observed quantitative conversion of ball-milled cellulose with remarkable hexitol volume productivity for conversion of concentrated cellulose feeds over a combination of heteropolyacids ($H_3PW_{12}O_{40}$, $H_4SiW_{12}O_{40}$) and Ru on carbon.

In an alternative approach by Van de Vyver *et al.* [7], Ni on carbon nanofiber catalysts (Ni/CNFs) are used for the catalytic hydrolytic hydrogenation of cellulose to sugar alcohols. While previous studies focus on cellulose cracking with expensive transition metals dispersed on porous supports, the concept of our catalyst design relies on the entanglement of threadlike carbon nanofibers (CNFs) around the water-insoluble cellulose matrix.

2.1 Catalyst synthesis

For the growth of CNFs, 20 wt% Ni/γ-Al_2O_3 was prepared by impregnating γ-Al_2O_3 (Condea Chemie; Puralox, 155 m^2/g) with an aqueous solution of $Ni(NO_3)_2.6H_2O$ (Alfa Aesar), subsequent drying at 353 K, calcination in static air at 873 K (heating rate 5 K/min) for 5 h, and then reduction under H_2 at 873 K for 1 h. Samples of 0.5 g catalyst precursor were placed in a quartz reactor and the CNFs were grown at 873 K under CH_4 flow (80 ml/min) for 3.5 h. The resulting catalyst is denoted as 3.0% Ni/CNF, where 3.0% represents the nominal weight value of Ni loaded on the CNFs.

2.2 Catalyst characterization

The CNFs grown over Ni supported on γ-Al_2O_3 were examined with SEM (Figure 8 a, b). Furthermore, transmission electron microscope (TEM) images were recorded on a Philips CM30-FEG electron microscope, coupled to a post-column GIF200 system for electron energy loss spectroscopy (EELS) (Figure 8 d-f). TEM image 8e reveals elongated pear-shaped Ni nanoclusters with an aspect ratio of approximately 1.6, on the tips of the CNFs.

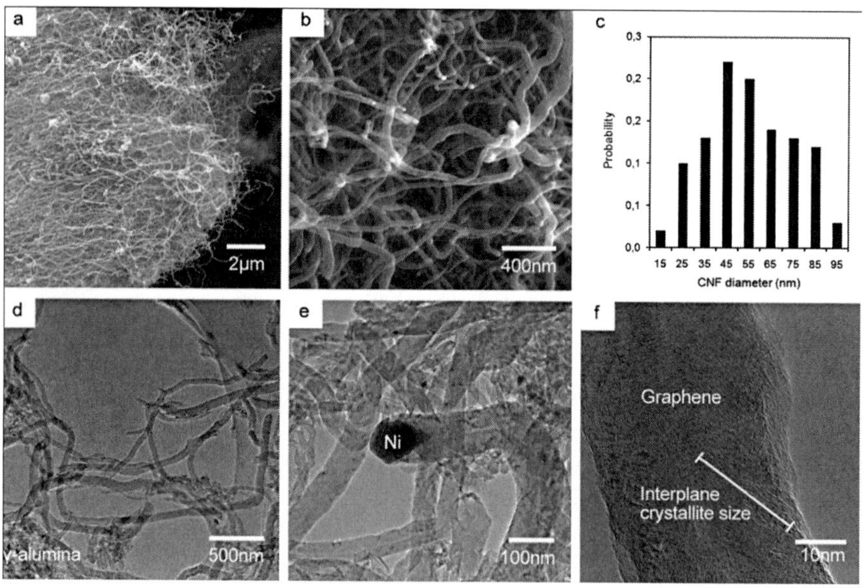

Figure 8: SEM images (a-b), CNF diameter distribution (c), and TEM images (d-f) of the 3.0% Ni/CNF catalyst sample.

Figure 9 below displays the powder X-ray diffraction (PXRD) patterns of the as-prepared 3.0% Ni/CNF catalyst. The identified diffraction peaks are indicated with an arrow. Clearly, for Ni nanoclusters located at the end of the CNFs, the intensity of the more exposed Ni(111) surface exceeds the reflections of the other crystal planes. Indeed, progress in understanding the growth mechanism of CNFs has demonstrated the reaction-induced reshaping of Ni nanocrystals, unveiling a larger proportion of the thermodynamically more stable Ni(111) surfaces.

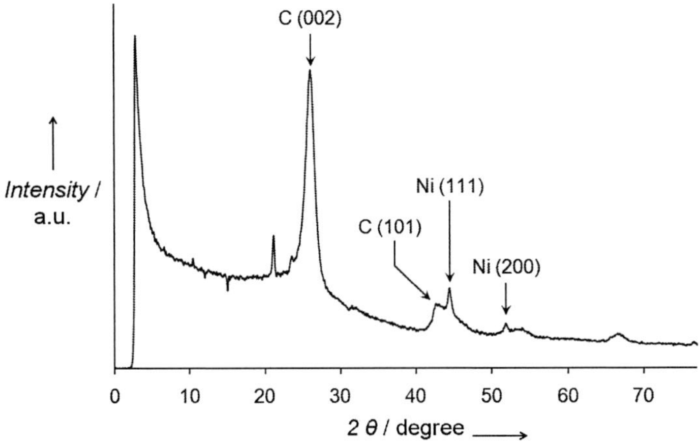

Figure 9: Powder X-ray diffraction pattern of the 3.0% Ni/CNF catalyst.

Nitrogen physisorption data were recorded with a Micromeritics Tristar 3000 apparatus at 77 K. Prior to physisorption measurements, the catalyst samples were evacuated at 473 K for 12 h. The N_2 adsorption isotherm of 3.0% Ni/CNF in Figure 10 can be classified as a type IV isotherm with a type B hysteresis loop. The agglomerated CNFs show a mesoporous structure, their Brunauer-Emmet-Teller (BET) specific surface area is measured to be 76 m^2g^{-1} and their total pore volume 0.16 cm^3g^{-1}.

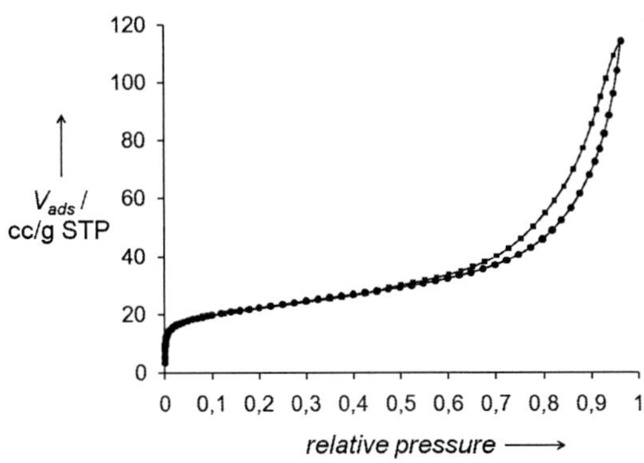

Figure 10: N_2 adsorption-desorption isotherm of 3.0% Ni/CNF.

Raman measurements were performed by putting the samples on top of a cleaned glass cover slip on an Olympus IX71 microscope. As excitation source a 633 nm wavelength laser was used. The laser light was reflected in the microscope on a 633 nm dichroic mirror towards a 100x 1.3 NA Olympus objective that focused the laser on the sample and collected the Raman signal. The Raman spectrum was recorded by using a Chromex Inc. Model 250 IS polychromator and a liquid N_2 cooled CCD camera. Raman spectra of the samples are shown in Figure 11. All the spectra have two typical peaks centered around 1330 cm^{-1} (D-mode) and 1590 cm^{-1} (G-mode). The integral intensity ratio I_D/I_G for the 3.0% Ni/CNF is 1.8, which corresponds to an in-plane graphitic crystallite size L_a of 21.4 nm.

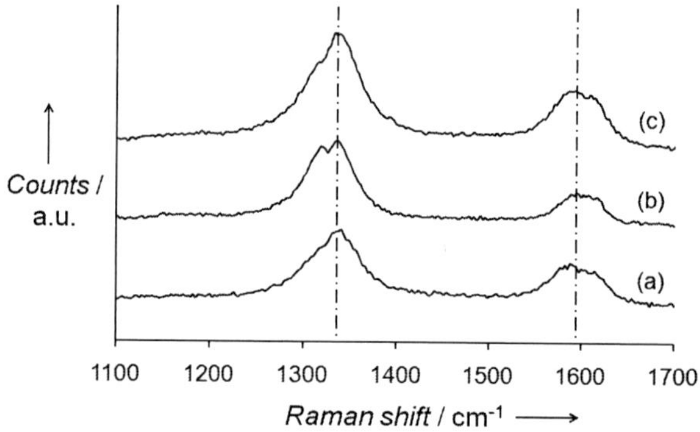

Figure 11: Raman characterization of the carbon nanofiber supported Ni:3.0% Ni/CNF (a), 6.1% Ni/CNF (b) and 3.8% Ni/CNF (c).

Nickel dispersions were determined using CO chemisorption (Figure 12). The Ni/CNF sample was loaded into a tubular reactor and pretreated in a stream of 20 mL/min of H_2 at 573 K (heating rate 5 K/min) for 1 h, and subsequently cooled to RT under a He flow of 20 mL/min. For the titration of the Ni surface, pulses of 5 μL of 100% CO were given with an interval of 2 min. The CO concentration in the outlet stream was followed with a quadrupole mass spectrometer.

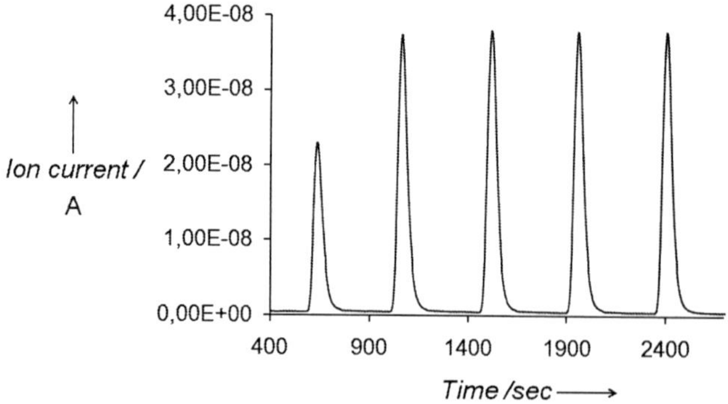

Figure 12: CO chemisorption analysis of the 3.0% Ni/CNF catalyst.

In the calculation of the dispersion, adsorption of 1 CO molecule per accessible Ni atom was assumed. The CO chemisorption analysis of 3.0% Ni/CNF yields a Ni dispersion of 0.17%.

2.3 Results and discussion

With these well-characterized Ni/CNF catalysts in hand, Van de Vyver *et al.* [7] demonstrated the production of sugar alcohols in excellent yields up to 70% sugar alcohols (sorbitol 51%, mannitol 6%, erythritol 13%) from ball-milled cellulose in a one-pot catalytic process. Sterically accessible Ni particles allow immediate hydrogenation of glucose, released after hydrolysis of cellulose. Despite the well-documented role of Ni as a hydrogenolysis catalyst, sorbitol appears to be surprisingly stable towards C-C and C-O bond breaking in the presence of the Ni/CNFs. In fact, the high hexitol selectivity of this catalyst is attributed to the structure-sensitive nature of the Ni-catalyzed reactions.

3 Conclusions and outlook

Selective conversion of cellulosic biomass is relevant to the development of catalytic strategies for production of valuable chemicals and renewable fuel components. The implementation of advanced concepts from heterogeneous catalysis for cleavage of the β-1,4-glycosidic linkages between cellulose anhydroglucose units represents a significant challenge for catalysis and is of major interest to our and other research groups. Conceptually, it is proven that specific design of solid catalysts even allows efficient conversion of this most recalcitrant family of renewable substrates, thereby bridging the gap between cellulose chemistry and heterogeneous catalysis.

In this respect, bifunctional Ni on carbon nanofiber catalysts and sulfonated silica/carbon nanocomposites present interesting approaches for direct catalytic transformation of cellulose into sugar alcohols and glucose, respectively, with the ability to control the cellulose conversion and product distribution. However, since both of the discussed catalytic reactions are still in their infancy, future research will be aimed at further optimization of reaction conditions to attain higher selectivities and greater tolerance of the catalysts towards dissolution and leaching. Work along these lines is currently in progress.

References

[1] Van de Vyver, S., Geboers, J., Jacobs, P.A. & Sels, B. F., Recent Advances in the Catalytic Conversion of Cellulose, *ChemCatChem*, **3(1)**, pp. 82–94, 2011.

[2] Lange, J.-P., Lignocellulose Conversion: An Introduction to Chemistry, Process and Economics (Chapter 2). *Catalysis for Renewables: From Feedstock to Energy Production*, eds. Centi, G. & van Santer, R.A., Wiley-VCH, Weinheim, pp. 21–51, 2007.

[3] Kamm, B., Production of Platform Chemicals and Synthesis Gas from Biomass, *Angew. Chem. Int. Ed.*, **46(27)**, pp. 5056–5058, 2007.
[4] European Technology Platform for Sustainable Chemistry, Industrial Biotechnology Section, 2005, http://www.suschem.org
[5] Fan, L.T., Gharpuray, M.M. & Lee, Y.-H., *Cellulose Hydrolysis*, Springer, Berlin, 1987.
[6] Van de Vyver, S., Peng, L., Geboers, J., Schepers, H., De Clippel, F., Gommes, C.J., Goderis, B., Jacobs P.A. & Sels, B.F., Sulfonated silica/carbon nanocomposites as novel catalysts for hydrolysis of cellulose to glucose, *Green Chemistry*, **12(9)**, pp. 1560–1563, 2010.
[7] Van de Vyver, S., Geboers, J., Dusselier, M., Schepers, H., Vosch, T., Zhang, L., Van Tendeloo, G., Jacobs, P.A. & Sels, B.F., Selective Bifunctional Catalytic Conversion of Cellulose over Reshaped Ni Particles at the Tip of Carbon Nanofibers, *ChemSusChem*, **3(6)**, pp. 698–701, 2010.
[8] Geboers, J., Van de Vyver, S., Carpentier, K., de Blochouse, K., Jacobs, P. & Sels, B., Efficient Hydrolytic Hydrogenation of Cellulose in Presence of Ru-loaded Zeolites and Trace Amounts of Mineral Acid, *Chem. Commun.*, accepted.
[9] Geboers, J., Van de Vyver, S., Carpentier, K., de Blochouse, K., Jacobs, P. & Sels, B., Efficient catalytic conversion of concentrated cellulose feeds to hexitols with heteropoly acids and Ru on carbon, *Chem. Commun.*, **46**, pp. 3577–3579, 2010.
[10] Zhao, H., Kwak, J. H., Wang, Y., Franz, J. A., White, J.M. & Holladay, J.E., Effects of crystallinity on dilute acid hydrolysis of cellulose by cellulose ball-milling study, Energy Fuels, 20, pp. 807–811, 2006.
[11] Rinaldi, R. & Schüth, F., Acid Hydrolysis of Cellulose as the Entry Point into Biorefinery Schemes, *ChemSusChem*, **2(12)**, pp. 1096–1107, 2009.
[12] Gommes, C.J. & Goderis, B., CONEX, a program for angular calibration and averaging of two-dimensional powder scattering patterns, *J. Appl. Crystallogr.*, **43**, pp. 352–355, 2010.
[13] Jacobs, P. & Hinnekens, H., Synfina SA, EP 0329923, 1989.

Interests and challenges of organic solvent nanofiltration for sustainable chemistry: the case of homogeneous catalysis of metathesis

M. Rabiller-Baudry[1], G. Nasser[1], D. Delaunay[1], A. Keraani[1], T. Renouard[1], D. Roizard[2], H. Ben Soltane[2], C. Fischmeister[1], J. L. Couturier[3] & D. Dhaler[4]

[1]*CNRS, Sciences Chimiques de Rennes, Université Rennes 1, France*
[2]*Laboratoire Réactions et Génie des Procédés, Nancy, France*
[3]*Arkema, Pierre Benite, France*
[4]*Novasep Process, Miribel, France*

Abstract

The development of chemical reactions catalysed by homogeneous organometallic complexes has allowed a significant breakthrough in synthesis chemistry, but now research axes must integrate catalysts recycling at the early stage of their conception. Catalyst recycling by nanofiltration in organic solvent (OSN) may represent an innovative route since this separation process at a molecular level is a low energy consumer. Challenges for OSN are membrane stability in organic solvents and the mastering of filtration conditions. Molecular engineering of catalysts based on slight structural modifications is also proposed to associate activity and high retention for OSN recycling. This paper deals with the integration of OSN within the homogeneous catalytic olefin metathesis reaction.
Keywords: nanofiltration, homogeneous catalysts, catalyst recycling, membrane material, olefin metathesis.

1 Introduction

Catalyzed transformations are recommended in the 12 principles of green chemistry as they can increase reactions efficiency and reduce the production of wastes by enabling selective and atom economy transformations [1]. However, it

doi:10.2495/CHEM110141

is quite clear that if the catalysts are not recyclable, the use of catalyzed reaction is less advantageous in terms of sustainable production. The recycling of catalysts is therefore an important scientific and technical challenge for fine chemistry. The most common approach, following the usual practices of organic synthesis, is to isolate the catalyst from the reaction medium at the end of reaction. Classically, extractions based on state changes (precipitation) are used that can often lead to more or less catalyst inactivation. To improve these separations several alternatives were considered:

- carry out the synthesis in non-conventional media [2], such as ionic liquids or supercritical carbon dioxide, perfluoroalkanes and water
- grafting catalysts on soluble supports in the synthesis medium but extractable by solvent change [3]
- grafting catalysts on insoluble supports [4], which facilitates recovery but is generally associated with poorer reaction kinetics and thus lower productivity.

All these solutions exist at the laboratory scale, but have drawbacks that have limited their transfer to industrial scale until now. The use of ionic liquids is a solution that has proven its effectiveness for the recycling of various catalysts, but have the disadvantage of not being eco-friendly solvent. Nowadays, chemists are looking for more reactive catalysts that could be used at "homeopathic" level and lost in the synthesis media. However, this prospect is not entirely satisfactory as the reaction products would be contaminated by metal residues. We put ourselves in a different perspective by proposing to separate the catalyst from the reaction medium using a membrane process (nanofiltration), which has the advantage of not requiring a change of state neither from the catalyst nor from the solvent. We thus expect a longer life cycle of the catalyst. Consequently, nanofiltration is an athermal process, which limits the energy required and its concomitant environmental impacts.

In this study, we apply this approach to olefin metathesis [5] which allows the efficient transformation of carbon-carbon double bonds. The three major contributors to olefin metathesis development and achievements shared the Nobel Prize in Chemistry in 2005 [6–8]. These reactions, usually conducted in toluene, can be catalyzed by homogeneous catalysts (soluble) containing ruthenium, as Grubbs or Hoveyda-type (figure 1) whose molecular weights are around 600 g mol^{-1} [9].

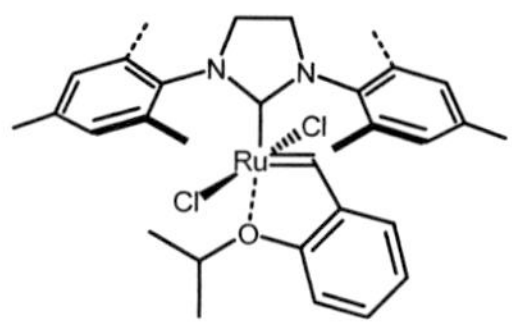

Figure 1: Hoveyda II catalyst (627 $g.mol^{-1}$) for olefin metathesis.

2 The challenges ahead for organic NF

Nanofiltration is a separation process at the molecular scale, widely used industrially in aqueous media and for which applications in organic solvent medium already exist for the recycling of solvents such as in the Max Dewax process [10]. Various studies at laboratory scale have been published dealing with the retention of homogeneous organometallic catalysts and recycling by organic nanofiltration (OSN) most often in methanol or other polar solvents: Co-Jacobsen (4 cycles) [11], Heck (5 cycles) and Suzuki (10 cycles) reaction catalysts [12, 13] and more rarely in an aromatic solvent (toluene) as for Hoveyda catalysts for the olefin metathesis [14]. However, to be widely used on an industrial scale, the NF in organic medium faces several challenges. The challenges are at different levels. The first bottleneck is the stability of membrane material in the solvent used for OSN which is also that of the organometallic synthesis. Obviously in practice inorganic membranes (zirconia, titania + zirconia on alumina support) can be selected at start as promising solvent resistant materials; but their pore size is generally too large to ensure a high enough level of retention of molecules less than 2000 g mol^{-1}, which will be the case of homogeneous catalysts. Hence actually the attention is focused on solvent resistant polymeric membranes, but this class of membranes is not yet well developed and there are only few commercially available polymer membranes that allow retention of smaller molecules. Basically, these membranes have composite structure having a tight skin layer on one side to ensure the retention and selectivity of separations. In these composite membranes, the skin layer is made of either polydimethylsiloxane (PDMS) or polyimide(s) deposited on a macroporous support (PAN, ...) chosen to provide mechanical strength to the overall structure. Contrary to aromatic polyimides, which are polar glassy materials, PDMS membranes, which are hydrophobic rubbery materials, do not exhibit the same stability and permeability in aromatic solvents such as toluene. For instance, PDMS composite structures can suffer from the separation of the skin layer from the macroporous support one after some time exposure to toluene due to different swelling of both polymers in toluene. Thus, there is a need to provide new membrane materials to develop more widely the OSN applications. Besides the solvent resistance, these membrane materials should ideally allow full retention of the catalyst. The development of such specific membranes is a long-term research. In the following, our research program proposes to develop two original routes. The first is based on an entirely homogeneous polymer membrane based on co-polyurethane materials. Conversely, the second one makes use of an organo-ceramic composite membrane obtained by the modification of a ceramic nanofiltration membrane by grafting of organo-titanate to improve the compatibilization of the membrane toward the organic medium. The second bottleneck is the mastering of the separation process itself. In aqueous media it is well known that retention performance depends on many physico-chemical and hydrodynamic aspects. Physico-chemical aspects takes place at different levels. First, the membrane fouling that is due to weak interactions between membrane and solute that adsorbs on it. Then solute-solute interactions that promote the establishment of deposit whose consequence is the permeate flux limitation and consequently a decline in productivity. The physico-chemical interactions are also involved in the transfer mechanisms of solute through the membrane:

attractive interactions (e.g. electrostatic or van der Waals type leading to adsorption on the membrane) generally favoured the transfer that is not only controlled by the relative size of the solute to the membrane pores. The intensity of these interactions is generally modulated by the filtration environment, for instance changes in retention should be expected by varying concentrations as well as impact of the overall physico-chemical environment. The hydrodynamics is involved in controlling the development of a more or less important fouling deposit and the phenomenon of concentration polarization upstream of the membrane. Concentration polarization is reversible and consists of an accumulation of solute on a few microns thickness. Both deposit and concentration polarization layer can be lowered by applying tangential filtration at high cross-flow velocity instead of dead-end filtration. First studies of the process mastering can be achieved with commercial membranes to draw trends but in fact the results will be dependent on the membrane/catalyst couple. Moreover, the membrane-solvent affinity controls the membrane swelling and thus opening of pseudo-pores in the polymer structure, thus allowing the occurrence of more or less convection inside the membrane in addition to diffusion controlled by the membrane/solute affinity. Consequently, for a given membrane and a given molecule the transfer mechanism can vary with the OSN solvent. Thanks to the lack of optimized membrane for the target separation, a complementary approach is to synthesize prototype catalysts which have good catalytic performances and improved retention with commercially available polyimide membranes. Considering the overall strategy, the OSN integration can be achieved in different ways according to the separation performance and to the metathesis conditions. Either OSN is made after the reaction performed in batch mode, according to a classical cascade: synthesis reactor then separation (figure 2), or OSN intervenes to set up a coupled process, which is then a synthesis membrane reactor (figure 3) that will be the most efficient regarding

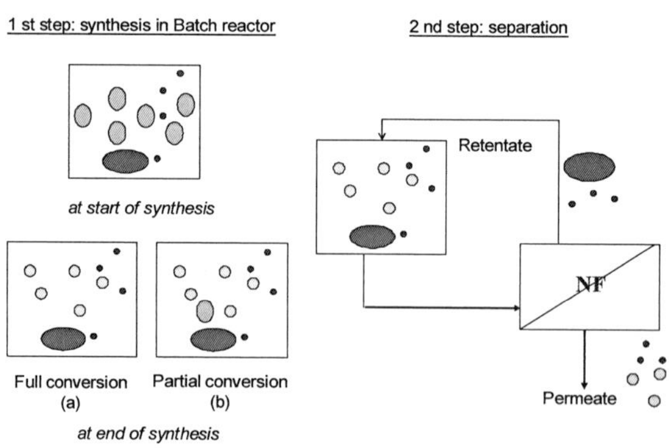

Figure 2: Integration of nanofiltration (NF) in synthesis process after reaction.

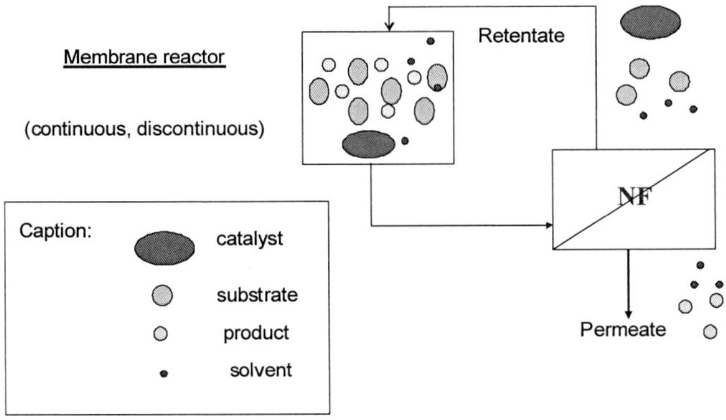

Figure 3: Integration of nanofiltration (NF) in a membrane reactor for synthesis.

sustainable requirements. There is no single answer. The choice depends on the reaction substrate and on the selectivity of the membrane towards the substrate and products of the metathesis. In the following we give some examples of the various approaches we have experienced.

3 Modifying catalysts to achieve high retention in OSN

We used pseudo-tangential filtration, meaning that tangential filtration was performed with very low cross-flow velocity ($v < 0.01$ cm.s^{-1}) actually corresponding to dead-end filtration. The membrane used was a polyimide one: Starmem 228 (MET, Evonik, UK). Its molecular weight cut-off (MWCO) is 280 g.mol^{-1}, which very roughly means that a 90% retention is expected for this molecular weight. Filtration is achieved until final volume reduction ratio VRR= 2, that is to say, with continuous extraction of permeate until halving the initial feed volume. At 25°C and 25 bar the commercially available Hoveyda II (figure 1) catalyst (0.53 mmol.L^{-1} in toluene) is 67% recovered, that is highly insufficient for recycling [14].

3.1 Synthesis and recovery of prototype catalysts

To increase the recovery of the catalyst, we have introduced modification in the Hoveyda II structure to increase its size by adding rigid phenyl groups decorated by sterically hindering substituents (figure 4). Recovery was actually increased up to 92% in standard conditions of OSN further used (figure 5). Nevertheless, the increase in the catalyst recovery was not only related to its size increase and to the increasing fouling as demonstrated by the "low" recovery of the heavy, bulky and rigid catalyst 5 (82%).. This underlines the important role of the affinity between membrane and phenyl groups of prototype catalysts leading to higher transfer through the membrane despite size increase [14].

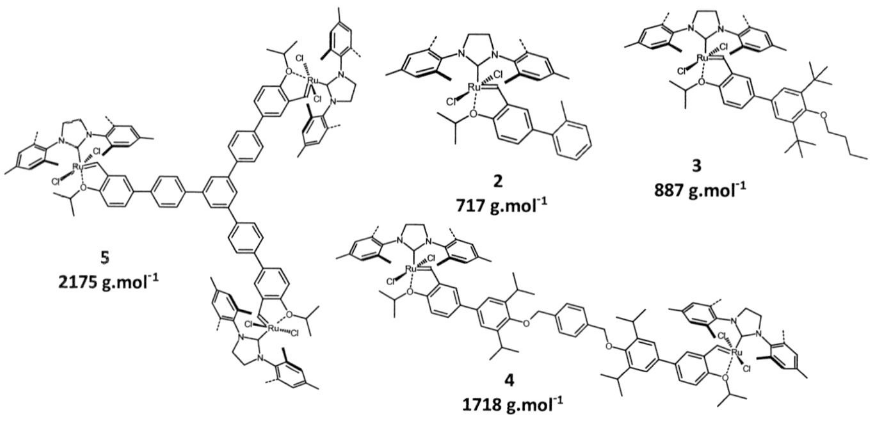

Figure 4: Prototype catalysts of Hoveyda II type.

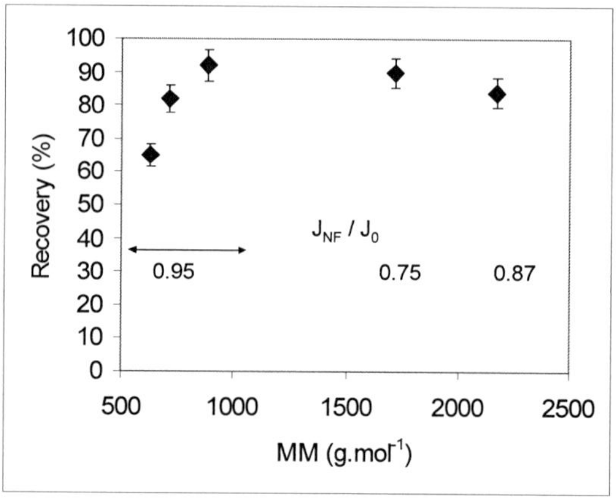

Figure 5: Recovery of Hoveyda II and prototype catalysts by OSN in toluene at VRR=2, 25 bar, room temperature, catalyst initial concentration: 2.5 to 3.1 mmol.L^{-1}, Starmem 228, pseudo-tangential (dead-end) filtration.

3.2 Recovery and recycling of prototype catalysts

As catalyst 3 (887 g.mol^{-1}) was the more retained, it was engaged in a series of model reaction of ring closing metathesis (RCM, figure 6). Each cycle was performed as follows. First, the metathesis reaction was achieved with 9 mg of catalyst 3 in 4 mL of solvent (toluene or dimethylcarbonate) and 100 mg of substrate (DATA). At the end of the reaction (monitored by gas chromatography) the reaction medium was diluted until a catalyst concentration of 2.6 mmol.L^{-1} (calculated from initial amount) was reached and OSN was then

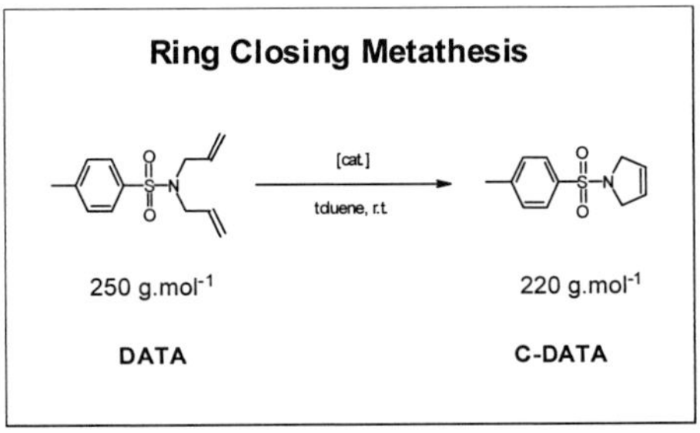

Figure 6: Model metathesis reaction.

performed in the standard conditions described before. The final retentate was then engaged in a further reaction with 100 mg DATA and so on. Up to 5 cycles were achieved with acceptable performances [14].

4 The control of transfers by the role of hydrodynamics

As the Starmem 228 membrane was no more commercially available, we change for the Starmem 122 (MWCO = 220 g.mol^{-1}, MET-Evonik, UK) also made of polyimide (not exactly the same as for the Starmem 228 as seen from FTIR spectra) and for the Duramem 300 (MWCO = 300 g.mol^{-1}, MET-Evonik, UK) that is cross-linked polyimide contrary to the two others. Actually, the MWCO of the 3 membranes are not significantly different. Hoveyda II catalyst was dissolved in toluene (0.53 ± 0.03 mmol.L^{-1}, 1/5 diluted compared with OSN with Starmem 228 membrane) and filtered at room temperature either with Duramem 300 in pseudo-tangential filtration (dead-end) or with Starmem 122 in tangential filtration (MET Cell, cross flow velocity close to 0.1 m.s^{-1}, as in spiral membrane). Pressure varied between 5 and 25 bar. The final VRR was between 1 (batch mode, total recycling of retentate and permeate) and 4. No variation of retentions was observed here by varying VRR. Figure 7 shows retention

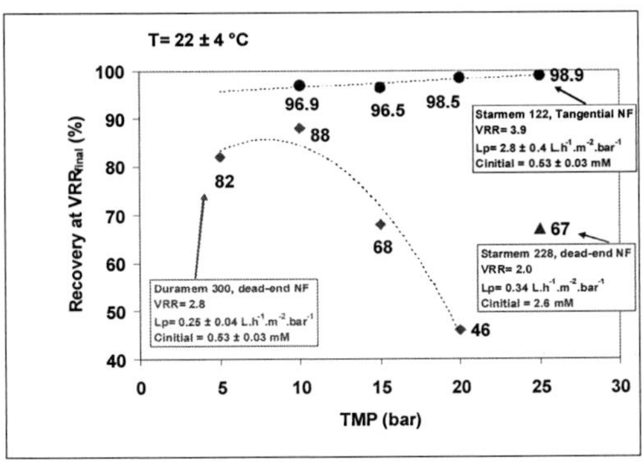

Figure 7: OSN of Hoveyda II commercial catalyst with different polyimide membranes and various hydrodynamics.

variation versus transmembrane pressure (TMP), clearly different from one membrane to the other. The most permeable membrane, namely Starmem 122 (Lp permeability was ten times that of Duramem 300) offers the better retention also thanks to tangential conditions.

5 The challenges of developing prototype membranes

It is important to be aware that even a small change in the membrane material composition can modify all the physico-chemical interactions in the filtration medium. Moreover, at constant chemical composition, slight change in the membrane porosity alters the transfer mechanisms and thus the selectivity.

5.1 Ceramic membranes modified by grafting of hydrophobic groups

The lower cut-off of nanofiltration (NF) ceramic membranes commercially available is MWCO = 1,000 g mol^{-1}. We tested such a membrane (Kerasep 1 $kg.mol^{-1}$, Novasep Process, France) in conditions close to dead-end filtration (tangential filtration with v close to 2.2 $cm.s^{-1}$) at very low and not optimised TMP = 1.5 bar. The retention of $0.030.10^{-3}$ $mol.L^{-1}$ Hoveyda II in toluene was less than 5% at VRR = 2. In addition, the membrane was strongly fouled as the permeate flux decreased by more than 1/3. To reduce the strong irreversible adsorption of the catalyst on the membrane we develop a surface modification by grafting of organo-titanate bearing highly hydrophobic functional groups (Figure 8). The chosen organo titanate was the isopropyl-isostearoyl-titanate. The retention of Hoveyda II catalyst (0.053 ± 0.001 $mmol.L^{-1}$) by the grafted membrane was increased up to 40% (2.2 $cm.s^{-1}$, 1.5 bar, VRR = 2, 20°C) These results are encouraging and must be improved in the future.

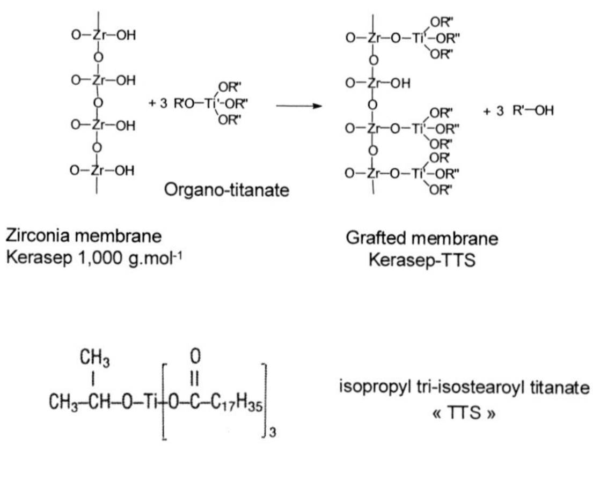

Figure 8: Scheme of zirconia membrane grafting according to [15].

www.witpress.com, ISSN 1743-3541 (on-line)

5.2 Polymer membranes

As said previously, the number of available commercial NF membranes is yet rather low and the related molecular cut-off are not so well predictable; moreover the NF properties of these membranes are known to vary with the applied NF conditions. Hence the synthesis of a series of original polymer membranes was carried out and the type of polymer network was chosen according the analysis of the metathesis reaction medium to be nanofiltrated as explained below. The targets to be reached by OSN membranes are very challenging and have been recalled in the above section. The first criterion to meet is the membrane chemical stability in the solvent under the OSN conditions, i.e. 8 to 15 bar at 40°C. Toluene and dimethylcarbonate (DMC) are the two solvents considered as the potential reaction medium of the metathesis reaction; they have quite distinct physical and chemical properties (table 1). Except the close molecular weights, toluene is a liquid having a low polarity whereas DMC is strongly polar as it can be seen from the solubility parameters. These distinct polar features are well reflected by their strong differences of solubility in water. Obviously it means that the polarity of the selected NF membrane will affect significantly the solvent OSN flux: as far as materials of similar glass transition are considered, a polar membrane will induce more affinities for DMC whereas a hydrophobic one will conversely have more affinities for toluene.

Table 1: Physical chemical properties of the solvent selected for the reaction.

	M ($g.mol^{-1}$)	Boiling point (°C)	Viscosity (mPa.s) 25°C	Solubility parameter δ ($MPa^{1/2}$)	Solubility in Water ($g.L^{-1}$)
Toluene	92	110.5	0.64	18.2	0.53
DMC	90	90	0.53	22.5	139

The second criterion to meet is the highest possible retention of the catalyst. Among the relevant parameters known to control the retention, some are linked to the catalyst characteristics, such as its molecular weight and its affinity for the membrane material. Others are also linked to the extent of the membrane interactions with the solvent: the stronger the interactions, the higher will be the membrane swelling, and the lower the catalyst retention. So the choice of the membrane polymer, either polar or non polar, will affect directly the retention efficiency in relation with the type of solvent used. To be able to control and tailor the polarity of the membrane and, thus, to adjust the membrane structure and properties, the synthesis of block co-polyurethanes (co-PU) was carried out. The reason of this choice is twofold. Firstly previous works have shown that this polymer structure is well stable at least in toluene up to 80°C [16]; secondly, the block co-polymerization route is rather a flexible one because it allows the easy incorporation of oligomer blocks, which can be used to tune the membrane polarity and to control the swelling extent of the network. The applied synthesis route follows the general polymerization reaction:

$$\text{Diisocyanate} + \alpha,\omega\text{-oligomer diol PEG} + \text{Triol} \rightarrow -[\text{D-PEG}]_{1-x} \text{---} [\text{D-T}_{2/3}]_x$$

$$\underset{1}{} \qquad \underset{(1-x)}{} \qquad \underset{2/3\,x}{}$$

with:

block D: -NHCO-R-NH-CO-, block P: -PEG-O-, block T: $T(\text{-O-})_3$

This scheme corresponds to the classical step polyaddition of difunctional reagents, i.e. diisocyanate and diols, under molar stoichiometric ratio; as diols, aliphatic polyethylene glycols (PEG) were used. The structures of the diisocyanate are given in the figure 9. To ensure chemical resistance, a third trifunctional reagent can be added as well; here we used a low molecular weight triol molecule. The polymer network was synthesis according to two successive steps: the initial formation of linear PEG urethane oligomers follow by the network extension and crosslinking. Thus the overall polymer network can be seen as the chemical association of two types of urethane units. Looking to the co-PU structure (Figure 10), it turns out that the polarity of the network can be varied by adjusting several keys parameters: (i) the PEG molecular weight, (ii) the diisocyanate molecular weight, and the extent of crosslinker versus PEG.

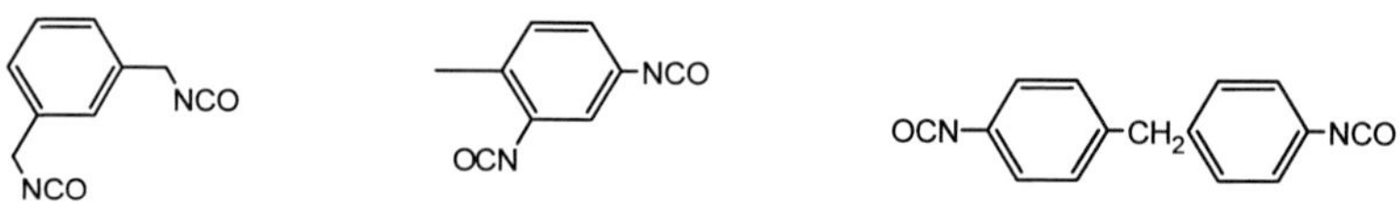

m-Xylylene diisocyanate | 2,4-toluene diisocyanate | 4,4'-Methylenediphenyl diisocyanate

Figure 9: Diisocyanate structures used to prepare block copolyurethanes.

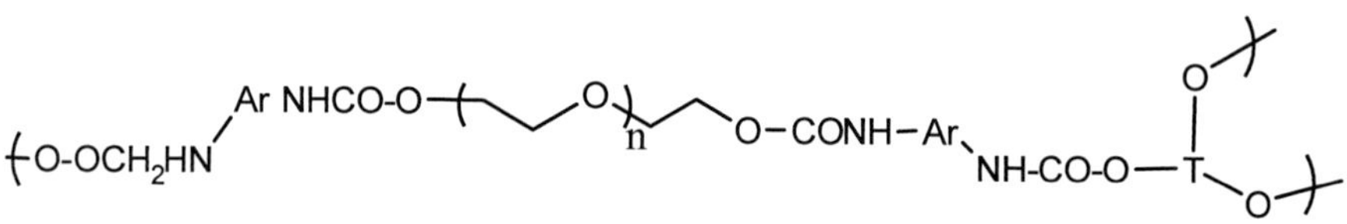

Figure 10: Chemically crosslinked block co-polyurethane-alcoxyether (PUE).

Accordingly to the above synthesis analysis, three membrane structures were first prepared to check the membrane feasibility; then the solvent compatibility was evaluated with a series of experimental data corresponding to immersion test in the pure liquid. The results gathered in table 2 showed a marked difference of swelling extent between the two solvents; going from toluene to DMC, the swelling increases from a factor 5 to 9.

Clearly the more polar solvent favored the swelling of PUE membranes. It can be explained by the polar PUE networks created from the aromatic isocyanates and the PEG of low molecular weight. Indeed, the aromatic urethane fraction is above 0.40 and so corresponds to a significant rigid, polar structure moiety compared to the PEG block. These preliminary results are in good agreement with the expected results and validate our synthesis membrane approach.

Table 2: Results of swelling experiments in organic liquids (t=20°C, P=1MPa).

Diisocyanate	Membranes	Aromatic fraction	δ*	Swelling extent (wt%) Toluene	DMC
TDI	PUE1	0.43	32.5	4.5	36
XDI	PUE2	0.45	26.4	8.5	40
MDI	PUE3	0.52	23.9	3.5	31

*Calculated for the equivalent aromatic urethane moiety (δ in $MPa^{1/2}$).

6 Conclusion

Our main long term goal is the integration of organic solvent nanofiltration within the homogeneous catalytic olefin metathesis reaction. It aims at achieving high retention of Hoveyda II type catalyst(s) to increase its(their) life cycle and decrease its(their) environmental impact due to loss at various steps. This example can be used as a model case to understand the various challenges facing fine chemistry nowadays and how integration of membrane separation processes can contribute to a solution that could be further scaled-up at industrial level.

The development of chemical reactions catalysed by homogeneous catalysts are recommended for sustainable production in fine chemistry according to the green chemistry concepts, but now, research axes must integrate catalysts recycling at the early stage of their conception, leading to an integrated approach including molecular engineering of catalyst together with recycling process. A dual objective has to be reached: (i) to increase catalyst productivity by increasing its recycle ability and simultaneously lower the environmental impact partly due to impossible re-use up today, (ii) to lower residual metal content in the reaction products and fulfil requirement for pharmaceutical uses, for instance. At laboratory scale, catalysts recycling by nanofiltration proved to be an innovative alternative route for recycling. This separation process is few energy consuming compared to other separation methods and the catalyst is not damaged thanks to the absence of phase change. Moreover, it is a friendly process toward fragile substrates and products that permeate through the membrane. Nanofiltration can be settled as a separation process or coupled to a reaction batch to lead to a membrane reactor. Membrane reactor is probably the best choice according to the sustainable requirements and to process intensification but not always well adapted depending on the target chemical reaction itself. Whatever the choice, separate separation or membrane reactor for synthesis, there are two main challenges for organic solvent nanofiltration. The first one is the membrane stability in organic solvents. Only few membranes able to highly retain homogeneous catalysts in polar, non polar or aromatic solvents are commercially available. Thus there is a need in preparation of new (selective) membranes to be able to widely extend OSN applications to various systems. We have shown here two types of membrane materials that can be developed for filtration in toluene that is one of the most difficult cases. One is an integrally polymeric membrane, the other is an organo-ceramic composite one. For both membranes, we only showed preliminary results that are promising.

Developments of these two membranes are still in progress. The second aspect concerned the mastering of the filtration process itself, needing a better fundamental understanding of transfer mechanisms not fully understood in OSN. Retentions are correlated to physico-chemical environment and hydrodynamics and synergies between these two types of parameters. For a given solvent/catalyst solution it was clearly shown how change in transmembrane pressure or filtration mode (dead-end or tangential) can strongly affect the catalyst recovery, in addition with the role played by the membrane material itself. Finally, in the future, we hope to be able to determine the best way for the eco-design of a whole process involving metathesis reactions in good agreement with the sustainable chemistry requirements.

Acknowledgements

The authors acknowledge the French National Research Agency (ANR) for financial support of the NanoRemCat2 project ANR-09-CP2D-11-01, the French Ministry of Research for AK PhD grant and KenReact (USA) for having kindly providing the organo-titanate reagents.

References

[1] Anastas P.T, Warner J.C., *Green chemistry : Theory and Practice*, Oxford University Press, New York, 1998.
[2] Cornils B., Herrmann W.A., Horvath I.T., Leitner W., Mecking S., Olivier-Bourbigou H., Vogt D., (eds), *Multiphase Homogeneous Catalysis*, 1st ed., Wiley-VCH Weinheim, 2005.
[3] Leadbeater N.E., Marco M., *Chem. Rev.*, **102**, pp3217-3274, 2002
[4] Song C.E., Lee S.G., *Chem. Rev.*, **102**, pp 3495, 2002.
[5] Grubbs R. H (Ed) *Handbook of Metathesis* Volume 1-3 Wiley-VCH, Weinheim, Germany, 2003.
[6] Chauvin Y., *Angew. Chem. Int. Ed.,* **45**, pp 3740-3747, 2006.
[7] Schrock R. R., *Angew. Chem. Int. Ed.,* **45**, pp 3748-3759, 2006.
[8] Grubbs R. H., *Angew. Chem. Int. Ed.* **45,** pp 3760-3769, 2006.
[9] Vogioukalakis G. C., Grubbs R. H., **110**, pp 1746-1787, 2010.
[10] White L.S., *J. Membrane Sci.*, **205**, pp 191-202, 2002.
[11] Aerts S., Weyten H., Buekenhoudt A., Gevers L. E. M., Vankelecom I. F. J., Jacobs P. A., *Chem. Commun.*, pp 710-711, 2004.
[12] Nair D., Scarpello J. T., White L. S., Freitas dos Santos L. M., Vankelecom I. F. J., Livingston A. G., *Tetrahedron Lett.*, **42**, pp 8219-8222, 2001.
[13] Datta A., Ebert K., Plenio H., *Organometallics*, **22**, pp 4685-4691, 2003.
[14] Keraani A, Renouard T., Fischmeister C., Bruneau C., Rabiller-Baudry M., *ChemSusChem*, **1**, pp 927-933, 2008.
[15] Chaufer B., Liou J.K, Bouguen A., Rabiller-Baudry M., Millesime L., "Matière inorganique de filtration modifiée par greffage d'organo-minéraux et son procédé de préparation", French patent n° WO9929402, 1999.
[16] Roizard D., Nilly A., Lochon P., *Sep. Purif. Technol.*, 22-23, pp 45-52, 2001.

Section 4
Multifunctional materials

Activated ceramic materials with deposition of photocatalytic titano-silicate micro-crystals

P. De Luca[1], A. Chiodo[1] & J. B. Nagy[2]
[1]*Department of Planning, University of Calabria, Italy*
[2]*Department of Chemical and Materials Engineering, University of Calabria, Italy*

Abstract

The aim of this work was to prepare materials for use in buildings depollution, formed by titanosilicate microcrystals immobilized on ceramic surfaces, combining their unique properties, including the photo-catalytic properties of titanosilicates with high mechanical strength, with the thermal and chemical stability of the ceramic materials. In particular, titanosilicates, ETAS-10, belonging to the family of ETS (Engelhard titanium silicate), were studied.
Keywords: titanosilicate, ETAS-10, ETS-10, ceramic material, ceramic support, microcrystal, zeolite, multifunctional material, hydrothermal reaction, photocatalytic material.

1 Introduction

Today's environmental problems are forcing the search to find new materials with low environmental impact capable of responding to the increasingly looming ecological requirements. Through the development of knowledge and technology, new materials used in building and construction have undergone a considerable evolution in response to new standards of quality and possessing new and more articulate properties [1].

The application of new techniques to ceramic materials is a broad experimental field and involves a wide range of possible applications and uses so that, through coupling them with other materials, such as zeolites, new characteristics to the traditional properties of ceramic materials can be added.

The various applications of zeolites derive directly from their unique and diverse properties as ion exchange, catalysis, adsorption, molecular sieving [2],

WIT Transactions on Ecology and the Environment, Vol 154, © 2011 WIT Press
www.witpress.com, ISSN 1743-3541 (on-line)
doi:10.2495/CHEM110151

but in addition to these, in recent years, their photocatalytic properties were increasingly taken into account, in the same way as for traditional photocalytic agents such as, for example, titanium oxide [3].

Among the improvements sought in recent years in zeolite photochemistry, new zeotype materials, with porous nature, called ETS (Engelhard Titanium Silicate) [4–6], are included due to their structural properties to make them functional on photocatalytic processes and then with potential applicability in the degradation of pollutants.

These titanosilicates microporous materials have recently attracted attention as photocatalysts for two particular aspects: the first is that the ETS are semiconductors containing photo-excited titanium fractions, the second is that their structure is characterized by a series of three-dimensional channels [7, 8]. The first aspect is certainly the most important in terms of active photocatalytic process for the formation of photo-induced torque (electron and hole), responsible for initiation of redox processes that lead to the degradation of pollutants. The second aspect, however, allows the titanosilicates to act as photo-selector form with respect to the pollutants molecules with which it comes into contact.

In order to modify the properties, aluminum has been incorporated into the structure of ETS-10 phase giving rise to a new type of microporous material called titanium-aluminum-silicate ETAS-10 [9].

Many researchers have studied the synthesis of ETAS-10 with different contents of aluminum in the structure. They have shown that the structure of this material is closely similar to that of ETS-10. The ETAS-10 features all the properties and application capabilities of the ETS-10 as catalysis, adsorption, ion exchange and photo-reactivity [10].

In this work the experimental conditions were sought to optimize the deposition of micro crystals of ETAS-10 phase on different ceramic types supports.

The ultimate aim was to investigate the possibility of creating de-polluting material usable for buildings, made of micro crystals immobilized on ceramic surfaces, which can combine the unique properties of zeolites, including those photo-catalytic ones, with the high mechanical strength, thermal and chemical stability of ceramic materials.

2 Experimental part

The experimentation was divided into distinct but subsequent events.

2.1 Preparation of different types of ceramic supports

In particular, different support types were prepared: (a) only clay, (b) clay and TiO_2 (Tioxide anatase), (c) clay and dried gel precursor of ETAS-10 phase (d) clay and crystalline ETAS-10. All ceramic supports were prepared using the following system:

100g. Clay : 35g. Distilled water : 10g. Additional phase

where additional phase=titanium oxide; dry gel precursor of the ETAS-10 phase; ETAS-10 crystalline phase.

The clay was pulverized by a pulverizer prior Fritsch Pulverisette model 2102 and then dried in an oven at 100°C for 1 hour. Titanium oxide, TiO_2, as commercial product (anatase, Alfa Aesar) was used. The dry gel and the crystalline phase were prepared using the following synthesis system expressed in moles:

$$1Na_2O: 0.6KF: 0.2TiO_2: 1.28HCl: 1.49SiO_2: 0.03Al_2O_3: 39.5H_2O$$

In particular, were prepared two solutions, one acidic and one basic. The acid solution was prepared by mixing HCl (Carlo Erba, 37wt%), $TiCl_4$ (Carlo Erba, 50wt%), $AlCl_3$*$6H_2O$(Merck) and distilled water. The basic solution was prepared by mixing KF (Merck, 40wt%), SiO_2 (Merck) and NaOH solution (BDH Analar, 50wt%). The two solutions were then mixed and a gel was obtained. Part of this gel was dried in an oven for 24 hours at a temperature of about 100 ° C and then pulverized, thus obtaining the powder of ETAS-10 dry gel, non-crystalline, to be included later in the support. The other part of the gel was poured into a cylindrical teflon, which in turn was inserted into an autoclave of stainless steel type Morey. Then everything was placed in an oven at 190 ° C for 72 hours after which the autoclaves were cooled under running water for about 15 minutes.

Figure 1: Shaping of ceramic supports in the slide for the X-ray diffraction.

The so prepared initial mixture, in order to achieve an adequate plasticity, was molded in the form of X-ray diffraction slides to be easily analyzed by XRD diffraction, fig.1.

The supports have been previously dried at 100°C for 24 hours and subsequently cooked in the oven at 600°C for two hours. The product was recovered by filtration and characterized by X-ray diffraction to ascertain the true nature of the phase and its crystallinity. The spectra obtained from XRD showed that the phase is ETAS-10 as it can be seen in fig. 2.

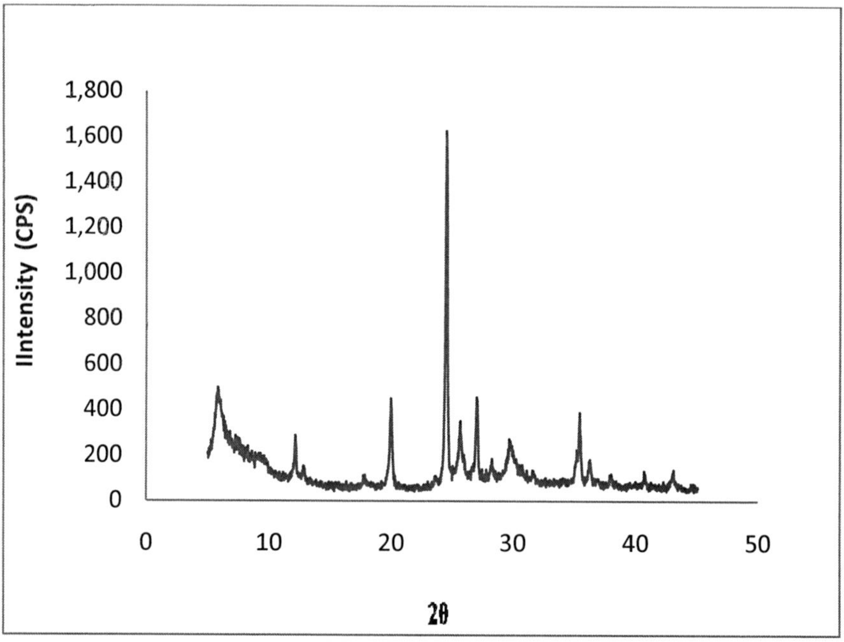

Figure 2: XRD patterns of ETAS-10 obtained at reaction temperature of 190°C and reaction time of 72 hours.

2.2 Activation of the surface of the ceramic supports

A part of the four different types of ceramic supports were subjected to a surface treatment, through a basic attack using a solution of 50 weight % of NaOH and leaving the supports in contact with it for 48 hours.

At the end a total of eight different supports were then obtained that have helped search for the best experimental conditions to obtain an optimal anchorage of the micro crystals on the surface of the supports as a function of the nature of added component and of the surface treatment.

2.3 Deposition of micro-crystals titanium silicate ETAS-10 through one or more cycles of hydrothermal reaction

The deposition of micro-crystals on the surface of the support was carried out through hydrothermal synthesis. In particular, each support was put in the precursor gel of ETAS-10 phase prepared as explained in Section 2.1, and the whole system was inserted in Morey type autoclaves and treated at a temperature of 190°C for 72 hours. After the expiry of the predetermined time, the autoclaves were cooled and then opened. The final products were thus: the support, a solution and micro-crystals deposited on the bottom of autoclaves, fig. 3. The support was then removed and washed with distilled water and dried at 100°C for 24 hours. The rest was filtered to recover the micro-crystals powder, and also they were washed and dried at 100°C for 24 hours. Subsequent XRD diffraction analysis was carried out both on the support and on the micro-crystal powder in order to verify that the phase obtained by initial gel was the same as that deposited on the support. On the same support were carried out several cycles of hydrothermal reaction in order to improve the deposition.

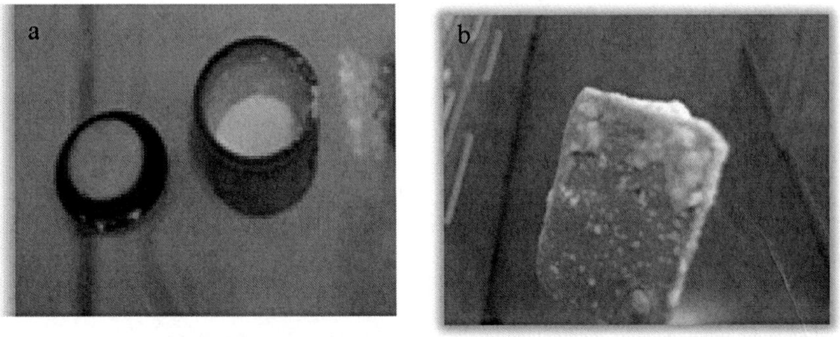

Figure 3: (a)Teflon container containing the final products. (b)Support extract before being washed.

2.4 Chemical-physical characterization of obtained samples

All samples were characterized by X-ray diffraction analysis (model Philips PW 1830) with Cu Kα radiation, the scanning speed was 0.02° per second and a range of between 5 and 45 2Theta.

In order to allow a comparison between the crystallinity of various samples, the XRD spectrum of ETAS-10 phase powder obtained after 72 hours of reaction at 190 °C was taken as reference. To do this, two characteristic peaks, in particular, those at 20.05 and 24.57 2Theta and their respective relative and peak intensities were taken as a reference, fig.2. Applying the following expression:

$$\text{Crystallinity} = [(I_{peak})_1 * (I_{rel})_1 + (I_{peak})_2 * (I_{rel})_2]/2$$

a value of crystallinity was obtained which has been assigned a value of 100. Using the same procedure, the percentages were calculated for all crystalline samples. Scanning electron microscopy SEM Stereoscan (360S) was used to determine the best conditions for the crystals growth on the ceramic surface, depending on the support type, the number of hydrothermal reaction cycles and surface pretreatments. Moreover, the compressive strength of the samples was measured to study the influence of hydrothermal treatments on the mechanical properties of the ceramic support by hardness tester Vander Kamp VK200.

3 Results and discussion

All samples obtained showed a change in visual appearance of the surface as a function of the type of support, of the repeated cycles of reaction and of the surface treatment, fig. 4.

Support Type	Treatment	N° hydrothermal reaction cycles			
		0	I	II	III
Only clay	As made				
	NaOH (50wt%)				
Clay + TiO_2	As made				
	NaOH (50wt%)				
Clay + Gel dry ETAS-10	As made				
	NaOH (50wt%)				
Clay + Crystalline phase ETAS-10	As made				
	NaOH (50w%t)				

Figure 4: Samples obtained after repeated cycles of reaction depending on the type of substrate or surface treatment.

Images in fig. 4 show that the samples undergo, from first cycle to third cycle, a color change that becomes more evident in untreated support. The lighter color is logically independent of the firing temperature of the ceramic, since they are all cooked initially at the same temperature but probably this is due to reactions that occur during the hydrothermal reaction cycles, where the reaction gel has a basic pH around 12.

The results obtained by X-ray diffraction analysis were analyzed to understand the role played by the nature of the support, the number of cycles of hydrothermal reaction and surface treatment on the deposition of crystals of ETAS-10 phase.

The following figures 5-7 summarize a comparison of all experimental variables considered.

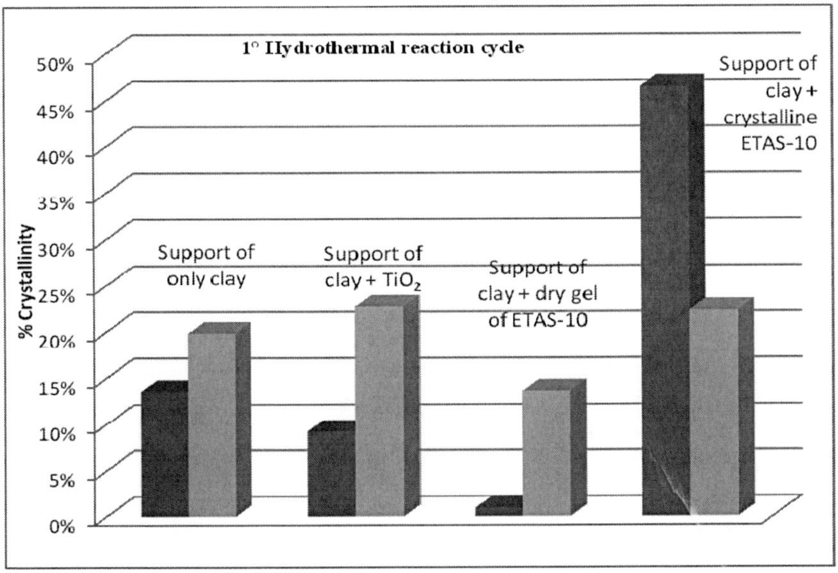

Figure 5: Crystallinity of samples obtained with one cycle of hydrothermal reaction as a function of support type and superficial treatment ■ Untreated; ■ Treated with NaOH.

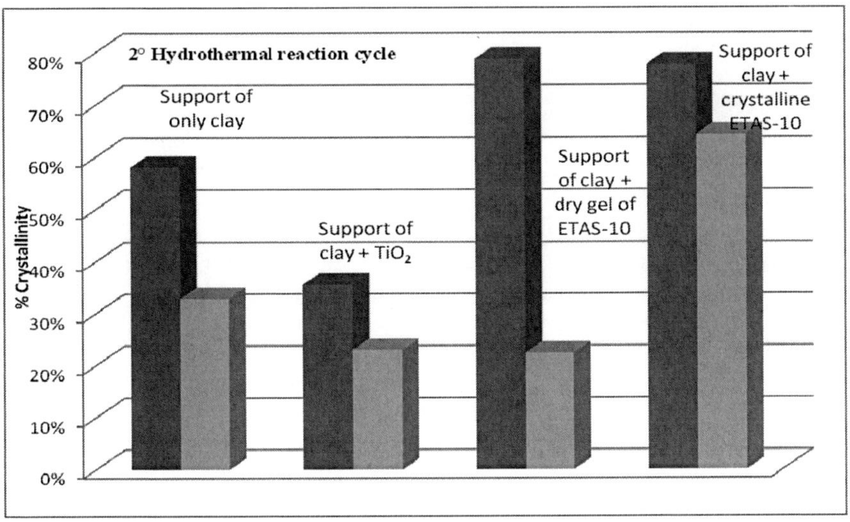

Figure 6: Crystallinity of samples obtained with two cycles of hydrothermal reaction as a function of support type and superficial treatment ■ Untreated; ■ Treated with NaOH.

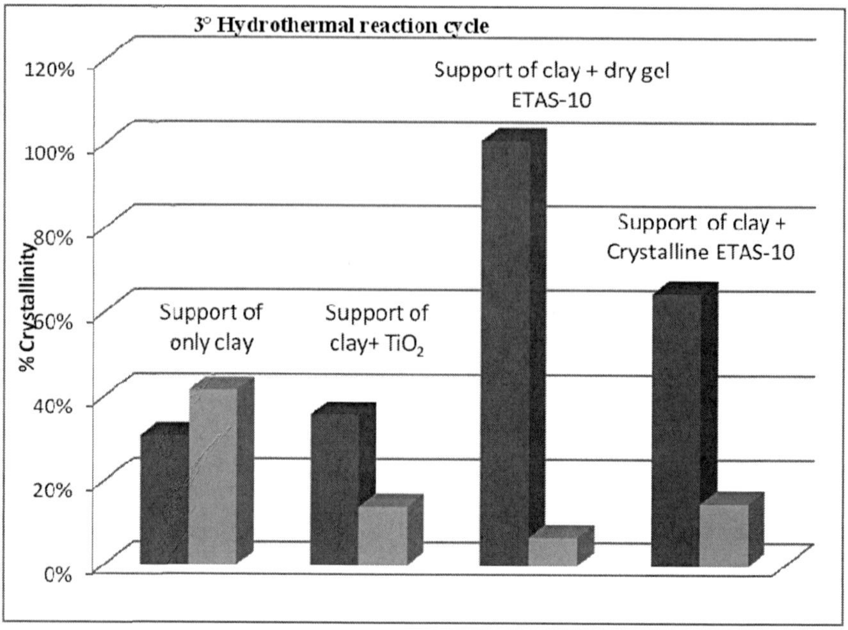

Figure 7: Crystallinity of samples obtained with three cycles of hydrothermal reaction as a function of support type and superficial treatment ▮ Untreated; ▮ Treated with NaOH.

Considering the preceding figures, the crystallinity of the different samples obtained, it is possible to make some observations.

Pretreatment with NaOH solution, improves the deposition only in the first reaction cycle if compared to that obtained on support that did not undergo pretreatment.

With the succession of reaction cycles, the basic activation of the substrate surface becomes less influential. This can be justified because the initial gel has an alkaline pH around 12 and therefore from the second reaction cycle all supports suffer the same basic attack. The type of support plays a key role.

It was possible to observe that the best crystalline deposition takes place on support prepared from clay and dry gel precursor phase ETAS-10. This is easily explained by the fact that the dry gel inserted into the ceramic substrate is able to provide the nutrients necessary for the formation of the first crystals in the right molar ratios. To this can be added the peculiarity that the gel is not only deposited on the support but it is an integral part thereof and thus it is capable of providing a potential accession to the initial growth of crystals.

The number of cycles affects the growth of micro-crystals on the support.

It was possible to observe that, for almost all samples, in the first cycle of reaction, the crystallinity, as evidenced by X-ray diffraction, remains fairly low, which indicates a low surface coverage. There are still obvious XRD peaks attributable to the support.

The second reaction cycle has an apparent lowering of the peaks (XRD) intensity of the support and an increase of peaks (XRD) due to the ETAS-10 phase thus confirming a good coverage of the substrate surface.

At the third reaction cycle there is a decrease of crystallinity but in some cases, only slight lowering occurs.

This could be justified by assuming a "saturation" of the surface of the support sites that are not able to trigger the growth of new crystallization nuclei and also to a partial detachment of micro-crystals, which were formed in previous cycles.

The best result was obtained on the support containing dry gel precursor of ETAS-10 phase, without pre-treatment with NaOH, and at the third cycle of synthesis.

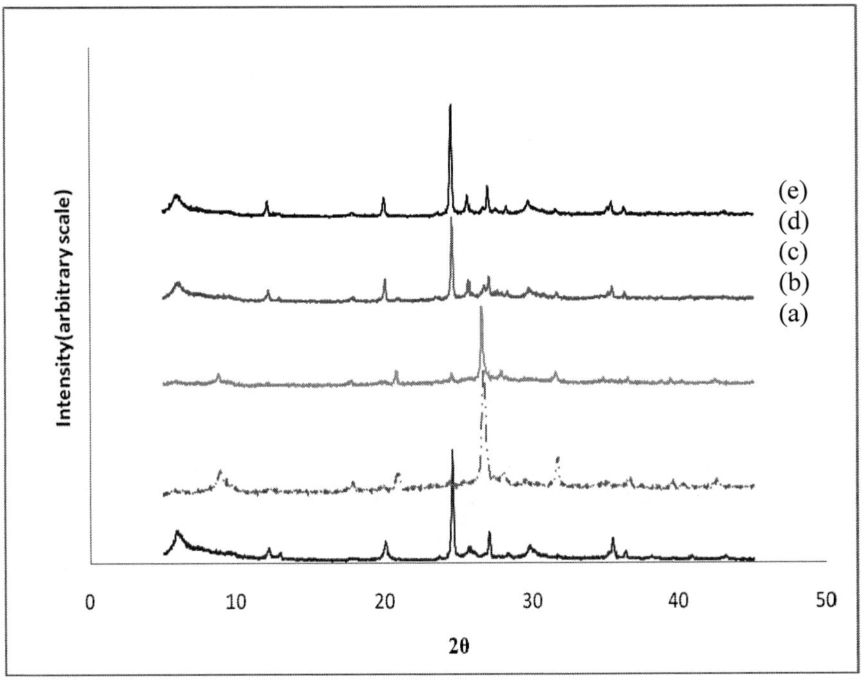

Figure 8: Comparison between XRD spectra of: (a) ETAS-10 powder; (b) support with clay and dry gel of ETAS-10 as made; (c) at first cycle; (d) at 2nd cycle; (e) at 3rd cycle.

The mechanical properties on all samples obtained were performed to verify if the different treatments, basic attack and hydrothermal reaction, affect their characteristics.

In the case of the treated support this threshold was consistently exceeded, although some effect of cleavage appeared on the supports treated and to the third cycle, revealing a strengthening of the support in contact with the NaOH solution.

Table 1: Compressive strength values of Kp (Kilopound) of the various supports. UT= untreated, T= treated with NaOH.

	Support Only clay		Support of Clay + TiO_2		Support of Clay + dry gel of ETAS-10		Support of Clay + crystalline ETAS-10	
	UT	T	UT	T	UT	T	UT	T
As made	20.6	>35	15.9	>35	>35	>35	22.0	>35
1 cycle	22.0	>35	17.5	>35	>35	>35	>35	>35
2 cycle	20.6	>35	>35	>35	>35	>35	>35	>35
3 cycle	25.2	>35	32.9	>35	>35	>35	>35	>35

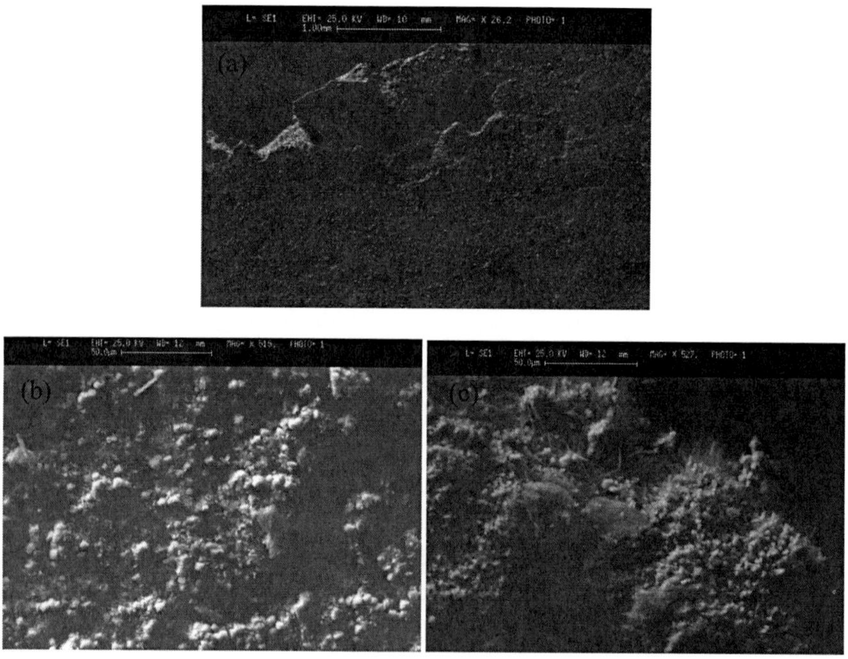

Figure 9: SEM micrographs of the support containing dry gel phase ETAS-10: (a) as made, (b) after 1 cycle of hydrothermal reaction (c) after the 2nd cycle of hydrothermal reaction.

The instrument used is calibrated to values not exceeding 35 Kp (kilopounds).

The same phenomenon happened on untreated support increasing the number of cycle so it could be inferred that, in general, as regards to mechanical resistance to compression, supports activated with NaOH showed the best results.

The SEM investigation has validated the data obtained through the diffractometric XRD study, highlighting the progressive growth of crystals with increasing hydrothermal treatment and a good anchorage of the crystals on the surface of the substrate.

4 Conclusion

The combined results obtained allow the following conclusions. The best experimental conditions for the deposition of micro-crystals of ETAS-10 phase on ceramic supports were obtained on support prepared with dry gel precursor phase ETAS-10 in the second reaction cycle hydrothermal without pre-basic treatment.

References

[1] Benedix, R., Dehn, F., Quaas, J., Orgass, M., Application of Titanium Dioxide Photocatalysis to Create Self-Cleaning Building Materials, *Lacer*, **5**, pp. 157-166, 2000.

[2] B. Nagy, J., Bodart, P., Hannus, I., Kiricsi, I., *Synthesis, characterization and use of zeolite microporous materials*, DecagenLtd, Szeged, Hungary, 1998.

[3] Dutta, P.K., Kim Y., *Photochemical process in zeolites: new developments*, pp. 483-489, 2004.

[4] Kuznicki, S.M., U.S. Pat. 4, 853,202 (1989).

[5] Kuznicki, S.M., Thrush, A.K., Eur. Pat. 0405978A1 (1990).

[6] Pavel, C.C., Vuono, D., Catanzaro, L., De Luca, P., Bilba, N., Nastro, A., B.Nagy, J., Synthesis and characterization of the microporous titanosilicates ETS-4 and ETS-10. *Microporous and Mesoporous Materials*, **56**, pp. 227-239, 2002.

[7] Shough, A.M., Doren, D.J., Nash, M., Lobo, R.F., Effect of Vanadium Substitution on the Structure and Photocatalytic Behaviour of ETS-10, *J. Phys. Chem C.*, **111(4)**, pp.1176-1782, 2007.

[8] Guan, G., Kida, T., Kusacabe, K., Kimura, K., Abe, E., Yoshude, A., Photocatalytic reactivity of noble metal-loaded ETS-4 zeolites, *Chem. Comm.*, **7**, pp. 618-620, 2004.

[9] Calza, P., Pazè, C., Pellizzetti, E., Zecchina, A., Shape-selective photocatalytic transformation of phenols in an aqueous medium. *Chemical Communication*, pp. 2130-2131, 2001.

[10] Liepold, A., Roos, K., Reschetilowski, W., Lin, Z., Rocha, J., Philippou , A., Anderson, M. W., Characterisation studies on the new microporous aluminiumcontaining ETS-10 molecular sieve used for processing larger molecules. *Microporous Materials*, **10**, pp. 211-224, 1997.

[11] Kim, S. D., Noh, S. H., Kim, Y. C., Hwang, J. Y., Jung, J. Y., Yi Sung Churl, Ki, W. J., Hydrothermal synthesis of aluminum-substituted titanosilicate, ETS-10, *J. Porous Materials*, **16**, pp. 307–314, 1997.

Developing a new certified reference material of brown algae for trace metal analysis

A. Santoro, M. Ricci, M. F. Tumba-Tshilumba & A. Held
European Commission- Joint Research Centre,
Institute for Reference Material and Measurements, Belgium

Abstract

A certified reference material (CRM) of brown algae is being developed at the Institute for Reference Materials and Measurements (IRMM), European Commission, Joint Research Centre, in order to supply academic and industrial analysts with a new tool useful in the environmental analysis of trace metals.

Sixty kg of raw brown algae (species *Fucus vesiculosus*) was collected from two different sites in Galway (Ireland) and processed according to ISO 34 and ISO 35. A large amount of milled and sieved (at <125 μm) algae material was bottled and subjected to detailed homogeneity and stability tests with regard to the parameters to be certified.

The between bottle homogeneity of the total element content of As, Cd, Cu, Hg, Pb, Se and Zn was assessed via a dedicated test, which confirmed the suitable homogeneity of the candidate CRM. The short term stability of the material when stored at high temperature (60°C) for up to 4 weeks was also considered, confirming no degradation under the conditions applied.

Milled and sieved algae was also checked for water content by using Karl-Fischer titration and several oven temperatures (from 85°C to 110°C) in order to establish the most reliable water determination method for the material.

Along with these studies, a long term stability isochronous study and the characterisation of the total element content for the above-mentioned metals by using an inter-laboratory comparison approach are in progress, so as to finalise production and release onto the market the brown algae CRM for trace metal analysis.

Keywords: brown algae, trace metals, certified reference material, fucus vesiculosus, environmental analysis.

doi:10.2495/CHEM110161

1 Introduction

In recent decades, marine pollution has been evaluated and monitored not only by analysing water and sediment, but also by using aquatic bioindicator organisms. Aquatic organisms, such as mussels and algae, have displayed the capacity to accumulate organic and inorganic pollutants; and algae, in particular, showed a significant uptake capacity for trace metals.

Among the three main groups of algae (brown, red and green), brown algae seem to have the highest biosorbent capacity, especially for Cd, Cu, Zn, Pb, Cr and Hg in water and wastewater [1, 2]. Specifically, *Fucus vesiculosus* and *Ascophillum nodosus* have been increasingly used in water quality monitoring [3].

In this context, the Water Framework Directive [4], although not providing environmental quality standards for trace metals in brown algae, suggests the use of macroalgae as an indicator of water quality. Thus, a number of European countries (mainly Finland, Sweden, Germany and Great Britain) have used macroalgae (and in particular *Fucus vesiculosus*) as a marine bioindicator [5, 6].

Furthermore, *Fucus vesiculosus* is also used in food, cosmetic and clinical fields [7, 8] for its nutritional and therapeutic properties; increasing in this way the interest of the scientific community in monitoring the level of trace metals.

However, in trace metal monitoring, either in the environmental field or food market, the possibility of performing the analyses in a reliable and accurate way is of crucial importance. To this end, certified reference materials (CRMs) are developed with the specific scope of providing a common reference point useful for ensuring and improving the traceability of measurement results.

As specifically stated in ISO Guide 34 [9] and ISO Guide 35 [10], in order to be recognised as a reference material, special requirements have to be met. Furthermore, specific tests to prove its homogeneity and stability over a long period of time need to be performed.

By following these guidelines, raw Bladderwrack (species *Fucus vesiculosus*) was processed and tested at the Institute for Reference Material and Measurements (IRMM, Belgium), with the aim of producing a new CRM of brown algae for use in trace metals analysis.

2 Processing

Sixty kg of raw brown algae (Bladderwrack, *Fucus vesiculosus*) was collected at two sites in November 2009; Silver Strand beach and Spiddal in Galway (IRL). Plants were cut above the holdfast by using a sharp knife, immediately rinsed with seawater to remove debris and sand and then collected in bags. In the laboratory, plants were checked for large epiphytes and other animal material, quickly rinsed in freshwater and immediately frozen at - 20°C. The collection was performed taking into account the seasonal variation of element content in algae [11]. Several authors have reported an increase in the content of some trace elements between January and February; whereas a significant decrease was observed during the summer period (July, August) [12].

Although the temporal variation of metals in algae has been differently interpreted in literature [13, 14], in our case, the algae were collected in November, so as to reduce the risk of collecting raw material during a period of potential seasonal instability.

The frozen raw material was transported to the IRMM (Belgium) in order to be processed under controlled humidity and temperature conditions.

Algae were washed with deionised water to eliminate sand excesses, dried in a drying cabinet (Elbanton, NL) at 25 ± 5°C and milled by using a cryo-grinding vibrating mill (Palla mill, KHD, Humboldt-Wedag, Köln, DE) cooled at about -190°C. One of the main parameters which can seriously affect the quality of a reference material is the homogeneity; potential sources of heterogeneity were reduced by accurate milling and sieving steps. The material was sieved at 125 µm and finally mixed for several hours by using a DynaMix CM200 (WAB, Basel, CH).

Particle size distribution was checked by using a Sympatec Helos (Clausthal-Zellerfeld, DE) from the beginning to the end of the processing and as presented in Figure 1, the resulting material showed an average particle size of ca. 100 µm.

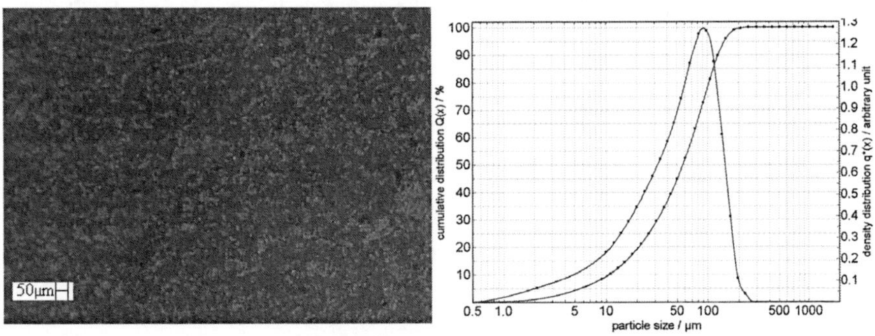

Figure 1: Typical micrograph for milled algae (left) and average particle size distribution obtained using isopropanol as dispersant (3 replicates, right).

During the whole processing, a loss of mass of ca. 85% was evidenced and the final portion recovered (with an average particle size of <125 µm) corresponded to ca. 14% of the total raw material mass.

Water content and activity of the material are key points for the stability of a CRM over long periods of time. Moisture contents above 5% can promote the growth of mould or enzyme activities with possible damage to the material [15]. Thus, the moisture level was decreased to 2.5% by vacuum drying the milled algae before bottling.

A large batch of about 1200 units was produced.

Possible causes of instability for the material generated during the processing were minimised by consistently working under controlled temperature and humidity and by filling the bottles in a glove box with a level of oxygen not exceeding 5%. The degradation due to UV radiation was minimised by using amber glass bottles, which were later double sealed in aluminium pouches previously flushed with argon.

Finally, the bottles were sterilised by gamma irradiation with a maximum radiation dose of 15 kGy, in order to minimise any remaining bacterial activity.

3 Water content determination

Water content determination plays a significant role in evaluating uncertainties of measurements, as element mass fraction is usually reported dry-mass corrected [16]. Water content can be significantly different according to the method determination used. Oven drying for instance, though being a commonly used method, tends to underestimate the values; while Karl-Fischer titration is more selective for water particles, but less used because of its high costs.

It is therefore essential to establish the best conditions of determining water content in the material which should combine suitability for a large number of users and at the same time reliability.

Milled and sieved algae was checked for water content by using Karl-Fischer titration (765 KF Coulometer Methohm, CH) and five different oven temperatures (85°C, 95°C, 100°C, 105°C and 110°C) in order to establish the most reliable water determination method for the material [17].

As shown in Figure 2, the results obtained demonstrate a consistent increase in the apparent water content by increasing the oven temperature until 105°C. Above 105°C, the water content seems to decrease.

The value obtained with the Karl-Fischer titration was 10.0 ± 0.3 g/100 g (average of 6 replicates measurements ± standard deviation), which was not found to be statistically different (F-test, 95% confidence level) from the value found with oven drying determination by using a temperature of 105°C.

4 Homogeneity and stability tests

In order to perform specific homogeneity and stability tests, several bottles were selected from the whole batch produced, according to a randomly stratified scheme [18] and then stored and analysed in different conditions.

For the homogeneity study, 11 bottles were selected within the whole batch, digested by using HNO_3 and analysed for the total content of As, Cd, Cu, Hg, Pb, Se and Zn with ICP-SFMS (Inductively Coupled Plasma – Sector Field Mass Spectrometry) under repeatability conditions (i.e. in one analytical run).

The results evaluated by using one-way ANOVA, showed a standard deviation between bottles always below 5% (Figure 3), confirming the homogeneity of the material.

Regarding the stability test, an isochronous scheme was followed. Eight bottles were stored at a temperature of 60°C for 0, 1, 2 and 4 weeks (two bottles per time point), in order to simulate harsh environmental conditions. After the testing time, bottles were brought to a reference temperature of -20°C so as to "freeze" their state and later on analysed together under repeatability conditions.

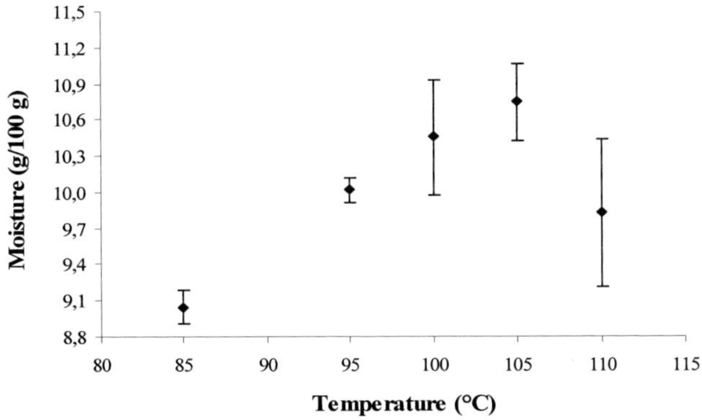

Figure 2: Water content (g/100 g) determination performed by using an oven drying method in brown algae (fucus vesiculosus). Bars correspond to standard deviation of 6 replicate measurements.

The results were obtained by digesting the samples with a mixture of HNO_3, H_2O_2 and HF and by using ICP-OES (Inductively Coupled Plasma – Optical Emission Spectrometry) for As, Cd, Cu, Pb and Zn and GF-AAS (Graphite Furnace – Atomic Absorption Spectrometry) for Hg and Se.

The selection of different digestion and measurement procedures was decided upon, in order to reduce the presence of biased results, related to problems arising from a specific method or procedure.

As for the homogeneity study, data were evaluated by using one-way ANOVA in order to detect any drifting of the elemental concentrations in the material over time. The slope of the linear regression was never found to be statistically significantly different from zero (Figure 4).

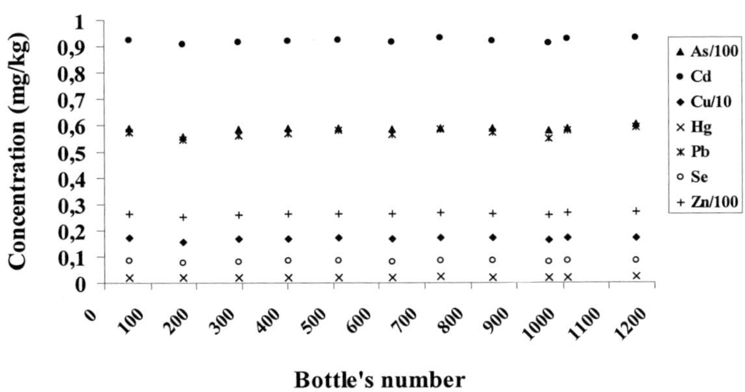

Figure 3: Homogeneity study performed on 11 bottles. Data are reported dry mass corrected in mg/kg. As and Zn are reported divided by a factor of 100 and Cu by a factor of 10.

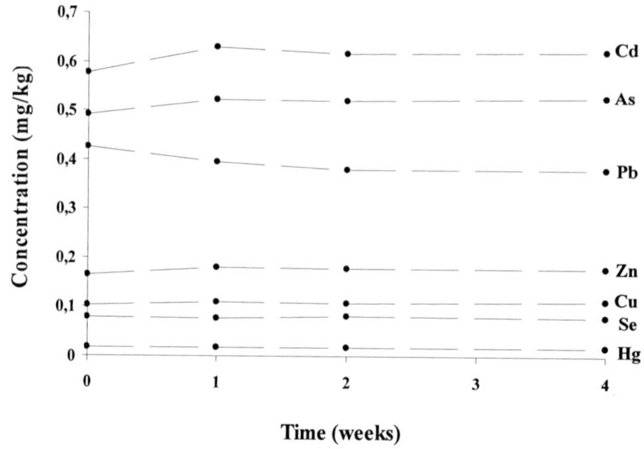

Figure 4: Stability study performed on 4 time points (0, 1, 2 and 4 weeks) at 60°C. Data are reported dry mass corrected in mg/kg. As and Zn are reported divided by a factor of 100 and Cu by a factor of 10.

Alongside this stability test, 14 bottles were stored, by using an isochronous study, for up to 12 months at a maximum temperature of 18°C, with the aim of evaluating the sample stability over a longer period of time and providing an estimation of the future shelf-life of the material.

A preliminary characterisation study on two elements (As and Zn) was performed by using neutron activation analysis (k_0-NAA) as a primary method [19]. The values obtained were in the range of 40-60 mg/kg for As and 20-30 mg/kg for Zn and thus comparable with the ones obtained during the homogeneity and stability tests (Figures 2 and 3).

A number of expert laboratories are currently being selected on the basis of specific quality criteria to take part in an inter-laboratory comparison campaign for the assignment of certified values for As, Cd, Cu, Hg, Pb, Se and Zn.

Key requirements in the selection are the use of different digestion methodology and analytical techniques, so as to minimise method dependant biases.

5 Conclusion

With the aim of producing a CRM for trace metal analysis in brown algae, several aspects concerning homogeneity and stability of the material, were checked. Homogeneity of the material was achieved during processing by employing an extensive milling and sieving procedure; specific preventive measures were taken to ensure stability, such as reducing the moisture content, bottling the material under an inert atmosphere and irradiating with a gamma source.

The homogeneity and stability tests performed so far show that the brown algae material is a good candidate for becoming a CRM.

Along with these studies, characterisation of the total element content, by using an inter-laboratory comparison approach, is in progress, so as to finalise production and release onto the market the CRM for trace metal analysis in brown algae.

References

[1] Davis, T.A., Volesky, B. & Mucci, A., A review of the biochemistry of heavy metal biosorption by brown algae-Review. *Water Research*, 37, pp. 4311-4330, 2003.

[2] Wang, J. & Chen, C., Biosorbents for heavy metals removal and their future. *Biotechnology Advances*, 27, pp. 195-226, 2009.

[3] Mata, Y. N., Blázque, M. L., Ballester, A., González, F. & Muñoz, J. A., Characterization of the biosorption of cadmium, lead and copper with the brown alga Fucus vesiculosus. *Journal of Hazardous Materials*, 158, pp. 316-323, 2008.

[4] European Commission, Directive 2000/60/EC of the European Parliament and of the Council establishing a framework for the Community action in the field of water policy, Off. J. Eur. Commun. L327.

[5] Amer, H. A., Ostapczuk, P. & Emons, H., Quality assurance in measuring the elemental composition of the alga Fucus vesiculosus. *Journal of Environmental Monitoring*, 1, pp. 97-102, 1999.

[6] Daka, E.R., Allen, J. R. & Hawkins, S. J., Heavy metal contamination in sediment and biomonitors from sites around the Isle of Man. *Marine Pollution Bulletin*, 46 (6), pp. 784-791, 2003.

[7] Teas, J., The consumption of seaweed as a protective factor in the etiology of breast cancer. *Medical Hypotheses*, 7(5), pp. 601-613, 1981.

[8] Burtin, P., Nutritional values of seaweeds. *Electronic Journal of Environmental, Agricultural and Food Chemistry*, 2 (4), pp. 498-503, 2003.

[9] ISO Guide 34, General requirements for the competence of reference material producers, 2000.

[10] ISO Guide 35, Reference materials - General and statistical principles for certification, 2006.

[11] O'Leary C. & Breen L., Seasonal variation of heavy metals in Mytilus edulis, Fucus vesiculosus and sediment from the Shannon estuary, *Biology and Environmental Proceedings of the Royal Irish Academy*, 98B(3), pp. 153-159, 1998.

[12] Riget, F., Johansen, P. & Asmund G., Natural seasonal variation of cadmium, copper, lead and zinc in brown seaweed. *Marine Pollution Bulletin*, 30(6), 409-413, 1995.

[13] Ströngren, T., Temperature-length growth strategies in the littoral alga Ascophyllum nodosum (L.). *Limnology and Oceanography*, 28(3), pp. 516-521, 1983.

[14] Villares, R., Puente, X. & Carballeira A., Seasonal variation and background levels of heavy metals in two green seaweeds. *Environmental Pollution*, 119, pp. 79-90, 2002.

[15] Igathinathane, C., Womac, A.R., Pordesimo, L.O. & Sokhansanj, S., Mold appearance and modeling on selected corn stover components during moisture sorption. *Bioresource Technology*, 99, pp. 6365-6371, 2008.
[16] Zeiller, E. Benetka, E., Koller, M. & Schorn, R., Dry mass determination: what role does it play in combined measurement uncertainty? *Accreditation and Quality Assurance*, 12, pp. 295-302, 2007.
[17] Vassileva, E. & Quetel, C.R., Influence of the correction for moisture content on the quality of the certification of cadmium, copper and lead mass fractions in rice. *Food Chemistry* 106, pp. 1485-1490, 2008.
[18] D. S. Yates, D.S. Moore & G. McCabe, The Practise of Statistics, W.H. Freeman & Co., NY, 2003
[19] Greenberg, R. R., Bode, P., Nadai Fernandes, E. A., Neutron activation analysis: A primary method of measurement. *Spectrochimica Acta Part B: Atomic Spectroscopy*, in press.

Section 5
Bio-based materials

Preparation and characterization of biodegradable films from keratinous wastes of the leather industry

L. Barbosa[1], J. Costa[2,3], C. Rocha[1], O. M. Freitas[2], A. Crispim[3], C. Delerue-Matos[2] & M. P. Gonçalves[1]
[1]*REQUIMTE, Faculdade de Engenharia, Universidade do Porto, Portugal*
[2]*REQUIMTE, Instituto Superior de Engenharia do Porto, Portugal*
[3]*CIETI, Instituto Superior de Engenharia do Porto, Portugal*

Abstract

With increasing demand for sustainable materials, keratin wastes have been regarded, in recent years, as renewable resources worthy of exploitation. Each year, the Portuguese leather industry discards a considerable amount of animal hair, which constitutes a troublesome waste product. In this study, in an attempt to find a potential utilization for this waste, the preparation of biodegradable films from bovine hair, by thermo-compression, was tested. Bovine hair formulations with 20, 30 and 40 wt % of plasticizer (glycerol, lactic acid) were pressed into films, at 147 kN and 120°C or 160°C for 4 minutes. The mechanical properties, colour, solubility and water sorption isotherms of the obtained films were assayed. All films were opaque and dark. Solubility was higher for films processed at higher temperature and with lactic acid as plasticizer. The Guggenheim-Anderson-de Boer (GAB) model gave a good fit to the experimental results for the moisture sorption isotherms. The stress at break, σ, and the Young's modulus, E, decreased and the strain at break, ε, increased with the addition of increasing amounts of plasticizer.
Keywords: bovine hair, keratin, film, thermo-compression, mechanical properties, solubility, water sorption isotherms.

doi:10.2495/CHEM110171

1 Introduction

Leather industry has been considered as one of the highly polluting industries with an inevitable negative environmental impact. Processing hides and skins and converting them into leathers is characterized by a huge amount of liquid and solid wastes [1].

Large–scale production systems are adopted for leather processing in clusters and therefore, the industry receives focus of environmentalists and society. Consequently tremendous pressure is exerted by various pollution regulatory organisms [2].

The production in the Portuguese Leather Industry is roughly 100 million square feet per year. The main production is cattle leather with about 80% of the total production, 15% for sheep, and 5% for goats, horses, buffaloes and others [3].

The introduction of cleaner technologies, in order to generate a minimum of residues is urgent. Several alternatives to the conventional process have been studied. The liming process with hair recovery allows the hair removal from the skin without any damage; moreover it decreases the organic load and the production of solid wastes, which can be used as by-products [4].

Bovine hair is composed of over 90% protein, the main component being keratin, a fibrous and insoluble protein highly cross-linked with numerous intermolecular disulphide and other bonds [5]. Several reports described the preparation of films by casting the reduced keratin solution after mixing with chemical crosslinkers such as ethylene glycol diglycidyl ether and glycerol diglycidyl ether [6–8].

A thermal processing is a different way to produce films and it has advantages compared with the previously one like this technique is simpler and environmentally friendly [9].

The aim of this work was to explore the possibility to use bovine hair wastes of Portuguese leather industry as a source of keratin for film production by thermo-compression. The hair waste containing keratin was ground and modified with glycerol or lactic acid, which acted as a plasticizer. The granulometry of the hair and the quantity of plasticizer were optimized. Films were prepared by pressing the modified keratin at typical polymer processing temperatures. The resulting films were tested to determine the effect of processing conditions and plasticizer addition on their water solubility and water sorption isotherms and on their mechanical properties.

2 Materials and methods

2.1 Preparation of bovine hair

Bovine hair samples were provided by the Portuguese industry Curtumes Aveneda. Hair was removed from the bovine skin using Hair-Saving process [10]. Most commercial systems for hair-save unhairing are based on immunisation. Immunisation can be achieved by using sodium hydroxide, lime or calcium hydroxide; it usually takes 1-1.5 hours. The developed work consists in the addition of commercial calcium hydroxide solution (1% (w/w), for 1 hour), allowing the removal of hair without damage, during the liming process in the leather industry.

After this processing, the hair was filtered, dried at 60°C and subsequently it was milled (ZM 200, Retsch, Germany) and sieved (AS 200, Retsch, Germany) for 30 minutes to a final particle size of less than 50 μm. This granulometry was selected because in previous studies it was observed that, for particle size less than 50 μm, the films were more flexible and less brittle [11].

Samples were characterized with respect to parameters such as humidity, percentage of organic matter and mineral content, fat content and Kjeldahl nitrogen.

2.2 Film preparation

Ground bovine hair was mixed with different ratios (20 - 40 wt%) of plasticizer (glycerol or lactic acid) in a mortar during ~30 minutes, at room temperature. These plasticizer ratios were chosen to compare with other studies in the literature on keratin films [12]. Following mixing, 1.5 g of sample was distributed evenly in a circular mold with 5 cm of diameter. The moulded sample was sandwiched between aluminium foil and pressed into film in a Carver Press Autofour/15P (model 25-12HC, Carver, Inc., Wabash, USA), at 147 kN and 120ºC or 160ºC for 4 minutes. These conditions were established based on preliminary tests. After pressing, the film was removed and allowed to cool till room temperature.

Film thickness was measured at five different points on each film using a thickness comparator (Absolute Digimatic Indicator, model ID-F150, Mitutoyo Co., Japan) with a resolution of 1 μm. Reported thickness values are means of the five measurements.

Samples for testing were cut from the films and conditioned at 23 ± 2ºC and 53% relative humidity for at least 48 h.

2.3 Colour

Film colour was determined by a Minolta colorimeter CR300 series (Tokyo, Japan) using the CIELab parameters; lightness (L*) and chromaticity parameters – a* (red – green) and b* (yellow – blue) – were measured. At least five independent measurements for each condition were tested at room temperature and humidity.

2.4 Solubility

The film solubility (%) in water was defined as the ratio of water-soluble solids, after 24 h immersion in water, to the initial solids content [8].

The specimens for this test were cut with an area of 1.5 × 1.5 cm^2 and, after conditioning, were weighed and placed in Erlenmeyer flasks with 20 ml of distilled water, which were sealed with parafilm and kept at 25ºC. After the assay, these films and original films were dried at 105°C to constant weight for determination of solids content. At least three replicates of each film were tested and film solubility was calculated as the average value of the measurements.

2.5 Water sorption isotherms

Water sorption isotherms were determined by the gravimetric method. Pre-dried samples with dimensions of 1.5 × 1.5 cm^2 were placed in desiccators with different relative humidity (RH), imposed by the use of saturated salt solutions (Table 1), for a_w ranging from 0.11 to 0.90. The experiment was carried out at 25°C. Values for the water activity of the salt solutions at 25 °C were obtained from the literature [13, 14].

Table 1: Saturated salt solutions used to control relative humidity in desiccators.

Salt	Chemical formula	Water activity, a_w, of the saturated salt solution (at 25°C)
Lithium chloride	LiCl	0.112
Magnesium chloride	$MgCl_2$	0.328
Potassium carbonate	K_2CO_3	0.432
Magnesium nitrate	$Mg(NO_3)_2$	0.529
Sodium bromide	NaBr	0.576
Strontium choride	SrCl	0.709
Sodium choride	NaCl	0.753
Barium choride	$BaCl_2$	0.902

The samples were weighed periodically, using a balance Sartorius BP211D (Sartorius AG, Germany) with a resolution of 0.01 mg, until they reached constant weight, after which their moisture content was determined by the gravimetric method. The Guggenheim-Anderson-de-Boer (GAB) model, eqn. (1) [15], was used to fit experimental sorption data:

$$Xe = \frac{Xm(C\,k\,aw)}{(1-k\,aw)(1-k\,aw+C\,k\,aw)} \tag{1}$$

where X_e is the equilibrium moisture content on a dry weight basis at the water activity a_w, X_m is the monolayer moisture content on a dry weight basis and represents the water content corresponding to saturation of all primary adsorption sites by one water molecule, C is the Guggenheim constant and represents the energy difference between the water molecules attached to primary sorption sites and those adsorbed to successive sorption layers and k is the corrective constant taking into account properties of multilayer molecules with respect to the bulk liquid.

GAB equation parameters were calculated using non-linear estimation, Quasi-Newton method, with the software StatSoft, Inc - STATISTICA, version 6.0.

2.6 Mechanical properties

Mechanical properties of the films were evaluated in a texture analyser (TA.XT2, Stable Micro Systems, Surrey, UK) equipped with tensile test attachments, using five replicates. The test specimens consisted of strips of uniform width and length (1.5×6.0 cm^2), corresponding to 60% of the recommended size (ASTM D882) [16].

The distance between the grips was 4.0 cm and the applied test speed was 0.1 cm/min, while force (N) and deformation (% strain) were recorded.

The Young's modulus, E, was calculated from the initial linear part of the stress-strain curve. Tensile strength was calculated as the stress at break (σ) of the specimen and elongation at break (ε) was expressed as the percent change of the original length of the specimen between the grips of the texture analyser. These tests were run immediately after removing the specimens from the desiccators, to minimize adsorption/desorption of water by them.

3 Results and discussion

For all the tested conditions, cohesive films were obtained with variable diameter (5.9–8.5 cm) and thickness (0.22–0.46 mm). These two parameters varied inversely, as expected. Higher diameters and lower thicknesses were observed for films plasticized with glycerol (Fig. 1). For both plasticizers used, higher diameters and lower thicknesses were observed when the protein/ plasticizer ratio or the processing temperature was increased.

3.1 Colour

All films were opaque and dark. However, films with lactic acid were lighter (higher L* values) than films with glycerol (Fig. 1).

Figure 1: Films obtained by thermo-compression of ground bovine hair mixed with lactic acid (left) or glycerol (right).

Colour parameters (L*, a* and b*) decreased when processing temperature increased (Fig. 2). No pronounced differences were observed by changing the amount of plasticizer added.

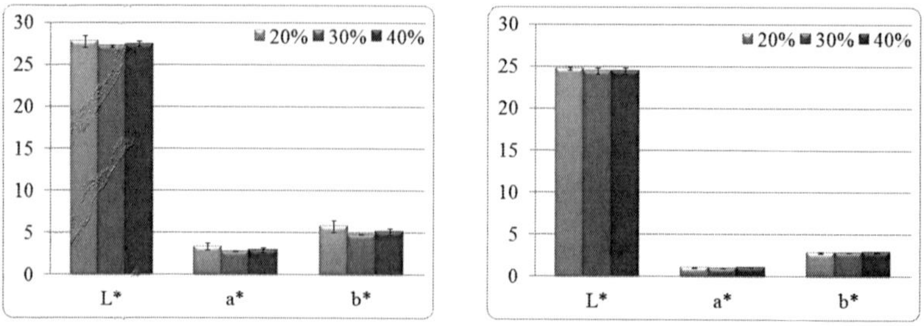

Figure 2: Colour parameters for films formed with glycerol at 120°C (left) and 160°C (right).

3.2 Solubility

Films solubility results are shown in table 2. In general, water solubility was higher for films processed at higher temperature and with lactic acid as plasticizer. The solubility increased also with the amount of added plasticizer (glycerol or lactic acid). Probably, the addition of plasticizer prevented the formation of -S-S- bridges on keratin films or interfered with other interactions like hydrogen bonding, electrostatic and hydrophobic bonds, resulting in an increase of solubility. Besides, both plasticizers are hydrophilic so it is expected that film solubility increases when their amount in the film formulation increases.

Table 2: Solubility of bovine hair films.

T (°C)	% Plasticizer	Solubility (%)	
		Glycerol	*Lactic acid*
120	20	8.6±2.9	15.5±0.1
	30	10.0±2.1	18.8±0.2
	40	13.2±3.4	20.6±0.2
160	20	13.6±0.7	17.4±0.2
	30	15.4±0.3	21.0±0.3
	40	28.6±0.1	25.5±0.1

The addition of glycerol produced a smoother, more homogeneous film surface than that produced with lactic acid which probably contributes for the lower solubility of those films.

The solubility values obtained in this study were lower than those reported in previous studies on keratin [8] or fish myofibrillar proteins [17] films produced by casting.

3.3 Water sorption isotherms

Moisture sorption isotherms for films with different plasticizer concentrations were obtained. In general, they presented a sigmoid shape, as shown in fig. 3,

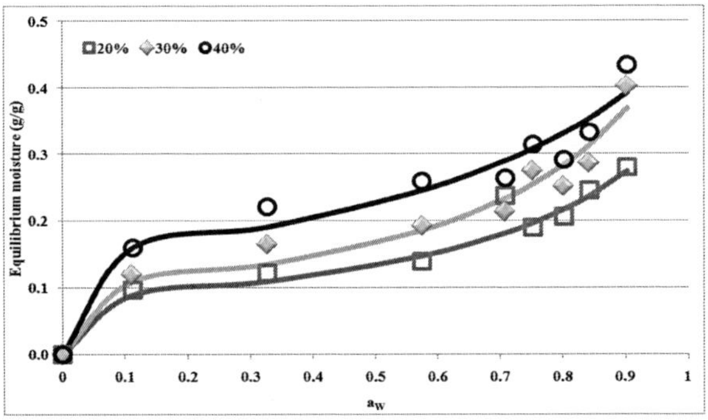

Figure 3: Water sorption isotherms for films processed with lactic acid, at 120°C. Lactic acid concentration (wt %): 20 (□), 30 (♦) and 40 (○).

indicating that the equilibrium moisture content, X_e, increased slowly with a_w up to 0.7, beyond which a steep rise in X_e was observed.

The GAB parameters and the correlation coefficients for films processed with lactic acid or glycerol, at different plasticizer concentrations and processing temperatures, are presented in table 3. Values of $C > 5.67$, $0.24 < k < 1$ and the correlation coefficient $R > 0.98$ show that the GAB equation gives a good fit to the experimental data.

In general, the moisture sorption isotherms were influenced by the concentration and type of plasticizer (Table 3). Higher levels of plasticizer increased the films water sorption (Fig. 3) and these results could be attributed to the hydrophilicity of the plasticizers, which presented hydroxyl groups capable of interacting with water by hydrogen bonds.

Table 3: GAB parameters for bovine hair films.

GAB Parameters		C	k	Xm	R	Temperature
Glycerol	20%	612502	0.947	0.075	0.989	120°C
	30%	1205028	0.892	0.093	0.974	
	40%	1732215	0.867	0.105	0.971	
Lactic acid	20%	2749025	0.778	0.081	0.995	
	30%	2069602	0.810	0.099	0.977	
	40%	496485	0.691	0.147	0.976	
Glycerol	20%	120	0.874	0.093	0.991	160°C
	30%	50	0.635	0.192	0.993	
	40%	71	0.761	0.141	0.941	
Lactic acid	20%	3944931	0.835	0.065	0.997	
	30%	45	0.718	0.104	0.921	
	40%	2500911	0.783	0.090	0.980	

3.4 Mechanical properties

The amount of plasticizer (glycerol or lactic acid) influenced the mechanical properties of the films. The stress at break, σ, and the Young's modulus, E, decreased and the strain at break, ε, increased with the addition of increasing amounts of plasticizer (Table 4 and Fig. 4). In other words, the films became more ductile with the addition of plasticizer (in the concentration range studied). This behaviour is in accordance with the results obtained for thermally processed keratin from poultry feathers [12].

Table 4: Mechanical properties for bovine hair films.

Plasticizer	% Plasticizer	T (ºC)	Stress at Break, σ (MPa)	Young's Modulus, E (MPa)	Strain at Break, ε (%)
glycerol	20%	120	4.43±1.20	9.51±0.49	0.49±0.14
		160	8.48±1.10	11.49±0.72	0.81±0.15
	30%	120	4.20±0.62	9.19±0.58	0.48±0.11
		160	3.82±0.35	3.53±0.78	2.12±0.26
	40%	120	3.78±0.91	8.91±0.70	0.44±0.07
		160	3.02±0.2	3.49±0.10	1.65±0.14
lactic acid	20%	120	6.22±0.37	10.47±0.14	0.60±0.05
		160	9.50±0.44	21.50±0.89	0.47±0.05
	30%	120	4.78±0.38	9.57±0.27	0.52±0.03
		160	6.74±0.72	11.32±0.19	0.61±0.08
	40%	120	2.89±0.23	7.82±0.08	0.39±0.03
		160	5.40±0.43	9.64±0.23	0.65±0.06

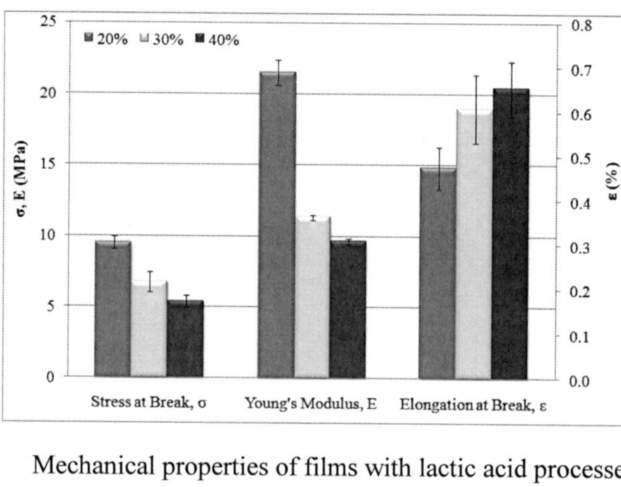

Figure 4: Mechanical properties of films with lactic acid processed at 160ºC.

Ductility also increased for films processed with glycerol (30% and 40%) when the processing temperature increased from 120ºC to 160ºC. The reverse was observed for films processed with 20% lactic acid that became more brittle.

Maximum values of σ = 9.50±0.44 MPa and E = 21.50±0.89 MPa were obtained for films with 20% lactic acid, processed at 160ºC (Fig. 4).

4 Conclusions

The results of the present work show that it is possible to obtain cohesive films from bovine hair by thermo-compression. This method is easy, quick and environmentally friendly because no solvents, oxidising or reducing agents are needed as often described in literature for the preparation of keratin films by other techniques such as casting.

The addition of a plasticizer at different levels modified the mechanical properties of the films meaning that it is possible to control the stiffness and rigidity of the films by varying the amount of plasticizer added. Films exhibited low water solubility when compared with other protein-based films described in the literature. This is an important result regarding the possible application of these films in food packaging or for agricultural uses.

Acknowledgements

Thanks are due to QREN (Quadro de Referência Estratégica Nacional), Fundo Europeu de Desenvolvimento Regional (EU) and AdI (Agência de Inovação) for financial support to FILMEQUE project: Development of keratin films from waste poultry feathers and bovine hair (SI IDT – 5551/2009 AdI) and for grants to L. Barbosa and J. Costa and to Curtumes Aveneda for providing the raw material.

References

[1] Kanagaraj, J., Velappan, K.C., Babu N.K.C. & Sadulla, S., Solid wastes generation in the leather industry and its utilization for cleaner environment - A review. *Journal of Scientific & Industrial Research*, **65**, pp. 541–548, 2006.

[2] Sundar, V.J., Gnanamani, A., Muralidharan, C., Chandrababu, N.K. & Mandal, A.B, Recovery and utilization of proteinous wastes of leather making: a review. *Reviews in Environmental Science and Biotechnology*, Springer, 2010. (DOI 10.1007/s11157-010-9223-6)

[3] APIC- Associação Portuguesa dos Industriais de Curtumes, Social & Environmental Report of the Portuguese Leather Industry, 2010.

[4] Ludvík, J., The scope for decreasing pollution load in leather processing, United Nations Industrial Development Organization (UNIDO), Buljan, J. (Project Manager), US/RAS/92/120/11-51, Regional Programme for

Pollution Control in the Tanning Industry in South-East Asia, Agro-Industries and Sectorial Support Branch, 2000.

[5] Lynch, M.H., O'Guin, W.M., Hardy, C., Mak, L., & Sun, T.T., Acidic and basic hair/nail (''hard'') keratins: their co-localization in upper cortical and cuticle cells of the human hair follicle and their relationship to ''soft'' keratins. *Journal of Cell Biology*, **103**, pp. 2593–2606, 1986.

[6] Tanabe, T., Okitsu, N. & Yamauchi, K., Fabrication and characterization of chemically crosslinked keratin films. *Materials Science & Engineering: C: Biomimetic and Supramolecular Systems*, **24**, pp. 441–446, 2004.

[7] Fujii, T., Ogiwara, D. & Arimoto, M., Convenient procedures for human hair protein films and properties of alkaline phosphatase incorporated in the film. *Biological & Pharmaceutical Bulletin*, **27**, pp. 89–93, 2004.

[8] Moore, G. R. P., Martelli, S. M., Gandolfo, C., Sobral, P. J. A. & Laurindo, J. B., Influence of the glycerol concentration on some physical properties of feather keratin films. *Food Hydrocolloids*, **20**, pp. 975–982, 2006.

[9] Barone, J.R & Arikan O., Composting and biodegradation of thermally processed feather keratin polymer. *Polymer Degradation and Stability*, **92**, pp. 859–867, 2007.

[10] Frendrup, W., Hair-Save Unhairing Methods in Leather Processing, United Nations Industrial Development Organization (UNIDO), Buljan, J. (Project Manager), US/RAS/92/120, Regional Programme for Pollution Control in the Tanning Industry in South-East Asia, Agro-Industries and Sectorial Support Branch, 2000.

[11] Pitrez, P.R., Costa, J., Rocha, C., Freitas, O.M., Crispim, A., Delerue-Matos, C. & Gonçalves, M.P., The effect of granulometry, glycerol concentration and presence of fat in the properties of films from feathers and bovine hair. *Macro2010- 43rd IUPAC World Polymer Congress*, Glasgow, UK, 11–16 July, 2010.

[12] Barone, J. R., Schmidt, W. F. & Liebner, C. F. E., Thermally Processed Keratin Films. *Journal of Applied Polymer Science*, **97**, pp. 1644–1651, 2005.

[13] Greenspan, L., Humidity fixed points of binary saturated aqueous solutions. *Journal of Research of the Natural Bureau of Standards*, **81A**, pp. 89–96, 1977.

[14] Spiess, W.E.L. & Wolf, W.R., The results of the COST 90 Project on water activity. In: Jowitt, R., (Ed.). *Physical properties of food*, Applied Science Publishers, London, pp. 65–87, 1983.

[15] Van der Berg, C., Description of water activity of foods for engineering purposes. In: B.M. McKenna (Ed.), *Engineering and food*, Elsevier Applied Science, New York, pp. 311–321, 1984.

[16] ASTM D882 – "Standard Test Method for Tensile Properties of Thin Plastic Sheeting", Annual Book of ASTM Standards, Vol 08.01, 2009.

[17] B. Cuq, N. Gontard, J-L. Cuq & S. Guilbert, Selected functional properties of fish myofibrillar protein-based films as affected by hydrophilic plasticizers. *Journal of Agricultural and Food Chemistry*, **45**, pp. 622–626, 1997.

Characterization of liquefied products from model woody components in the presence of mineral acid catalysts

Q. Wang[1], Q. Chen[1], P. Apaer[1], Q. Qian[1], T. Maezono[1], N. Mitsumura[1], H. Kurokawa[1] & X. Guo[2]
[1]*Department of Environmental Science and Technology, Graduate School of Science and Engineering, Saitama University, Japan*
[2]*School of Environmental and Chemical Engineering, Shanghai University, China*

Abstract

Cellulose and lignin are the main structural polymers in the plant cell wall. Cellulose is the structural component of the primary cell wall of green plants, many forms of algae and the oomycetes. About 40–50% of woody matter is cellulose. Lignin is a highly cross-linked polymer created by the polymerization of substituted phenolic compounds, known as monolignols, such as coniferyl, pcoumaryl, and synapyl alcohol. Liquefaction process is one of the promising techniques for effective utilization of woody biomass for the lignocelluloses can be converted to liquid reactive materials as the bio-based materials. Cellulose would have an advantage of providing liquefied product with small range of variance. The phenolated woody components have high acidity in the presence of mineral acid catalysts and possess the constituents which can react with formaldehyde. In addition, lignin, one of the major woody components including the hydroxyl-benzyl structure, has the potential to react with formaldehyde. However, as its complexity in structure, the liquefaction mechanism and the liquefied products with phenol should be found out to solve some problems such as the reaction efficiency and low molecular weight products, and it will be useful to preparation of bio-based materials thought the liquefaction processes. In our study, two model woody components have been used under the different liquefaction conditions with phenol. In our experiments, the model cellulose component is specially used in the experiment to test the characteristics of the

WIT Transactions on Ecology and the Environment, Vol 154, © 2011 WIT Press
www.witpress.com, ISSN 1743-3541 (on-line)
doi:10.2495/CHEM110181

products under different ratios. As the results, we found that the final liquefied products from model component substance of lignin only 17.2% (wt) and the growth rate is very low with the molecular weight (Mw) up to 2119 under the reaction temperature of 150°C and 3 hours in the liquefaction experiments. However, the model cellulose component was confirmed to contribute more. On the contrary, the Mw of raw woody powder material can be reach to 1851. In a series of the mixing experiments, we found that the variation of Mw in the different experimental conditions determined by a gel permeation chromatography. From the results of liquefaction residue, we calculated the activation energy, and compared the value of those in different conditions. The solubility of the phenolated woody powder had been evaluated in eight organic solvents to evaluate the hydrogen bonding strengths of these solvents. It is very helpful for sustainable chemistry if polymeric materials can be effectively produced from the biomass liquefaction processes.
Keywords: woody components, liquefaction, gel permeation chromatography, eco-polymeric materials, mineral acid catalysts.

1 Introduction

As one of the methods of using the biomass effectively, composition of the resin which used the biomass as materials was mentioned. It is necessary to liquefy the biomass and the technique of this liquefaction as a proceeding to produce the compounds of resin. At the early studies of liquefaction, in addition to organic solvents, such as phenol and alcohol, woody biomass was put into the airtight container, and it began from having succeeded in obtaining a viscous liquid by putting 250°C heat and pressure. Then, it succeeded in making pressure into normal pressure by adding an acid catalyst, and lowering reaction temperature to 150°C [1].

Lin used GG (*guaiacylglycerol-β-guaiacyl*), as a lignin model compound, carried out the relevant researches about liquefaction reaction mechanism of lignin compounds in the presence of sulfuric acidic catalyst [2, 3] and without catalysts [4, 5]. They had also attempted to clarify the reaction mechanism of cellulose with phenol under the acid-catalyzed conditions. It was found that the yield ratios of various compounds in the end reaction products are greatly dependent on the reaction conditions. Therefore, by controlling these reaction conditions, it is possible to adjust the structure and properties of the end liquefaction products [6]. Kobayashi *et al.* [7] reported that the wood contents of liquefied woody were limited to low levels, because the condensation reaction occurred concurrently among the wood components themselves during liquefaction. Recently, cellobiose, cellulose, and starch were used as model compounds, and their liquefaction mechanisms in the presence of polyhydric alcohol or phenol, with or without sulfuric acid as catalyst, were studied [8–10]. Sung *et al.* [11] reported the phenol liquefaction at the higher temperature of 150°C is considered to be via a predominant reaction pathway, mainly accompanying with the condensation reaction of the phenolic compounds produced during phenol liquefaction of cellulose.

In our study, model woody substances such as cellulose powder, alkaline lignin and their mixture were used for liquefaction processes to analyze the reaction mechanism of woody substances using phenol as an organic reactant. Woody powder was also liquefied as the comparison. Liquefaction efficiency and molecular weight distribution were evaluated from the liquefaction products.

2 Materials and methods

2.1 Materials

In our study, Japanese cedar (*Cryptomeria Japonica*), as the woody material, had been collected from the company of waste treatment factory. Cellulose (CAS No: 9004-34-6 MP Blomedicais, LLC) and lignin (Alkaline, CAS No: 9005-53-2 TCI. Co, Ltd) was bought on the market as model woody materials. The information of these model substances is listed in table 1.

Table 1: Characteristics of cellulose and lignin as model woody materials.

Item	Cellulose microcrystalline	Lignin(Alkaline)[a]
Formula	$(C_6H_{10}O_5)n$	$C_{10}H_{12}O_3$ $C_{11}H_{14}O_4$ $C_9H_{10}O_2$
Weight	(162.06)n	180.2 g/mole 210.2 g/mole 150.2 g/mole
Structure	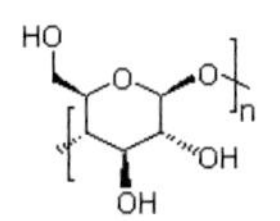	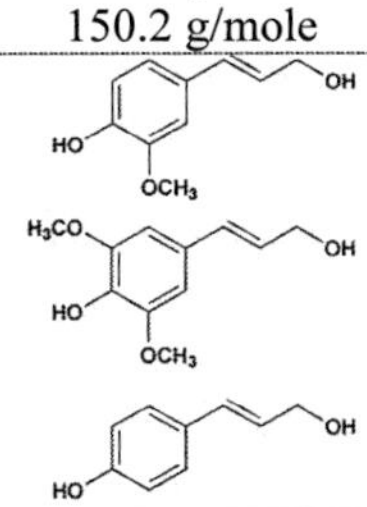

[a] The product is composed of three primary precursors coniferyl alcohol, mustard alcohol, p-coumaric alcohol dehydrogenase enzyme aggregates formed by plant polymers.

The oven-dried woody powder had been milled in smashing equipment and retained at the size ranges of 10-100 mesh screens were used for the test. All of these samples of woody materials were dehydrated in an oven at 105°C for 12 hours. The main different acidic catalysts such as H_2SO_4 (95%), H_3PO_4 (85%), phenol and other reagents were prepared with analytical grade reagents in accordance with the Japanese Industrial Standard (JIS) bought from Wako Pure Chemicals, Co. Ltd., Japan.

2.2 Chemical composition of liquefied woody materials

In the proximate and ultimate analysis, the ash content of the residue left after the combustion of woody sample is performed according to the industrial standard method (JIS-M8812). 7.7 wt% of the moisture of woody sample is determined by measuring the weight loss after drying the sample at 105°C in an oven. 79.1 wt% of volatile matter was tested by heating the woody powder to 900°C under carefully controlled conditions and measuring the weight loss. Then, 12.6 wt% of fixed-carbon in woody materials was calculated by subtracting the percentages of moisture, volatile matter, and ash of the woody samples. Other components were also quantitatively analyzed for the waste woody powder used in our experiments which were summarized in table 2.

Table 2: Composition analysis (wt%) of waste woody material.

	Composition analysis	Liquefied woody materials
Carbohydrate	Holocellulose	72
	Cross and Bevan cellulose	61
	Alpha Cellulose	52
	Klason Lignin	23
Solubility	1%NaOH	26.2
	EtOH/Benzene	3.3

2.3 Liquefaction experiments

Liquefaction experiments of waste woody materials and model substances were carried out by varying the weight charge ratio of wood to phenol with 1/4 and the reaction temperature at 110°C, 140°C and 170°C controlled by an oil bath. The raw materials put into a 500ml three necked flask equipped with stirrer. Reactants treated with methanol as the diluents. We added the catalyst to the reactant as the reaction start time, and added the methanol into liquefied products as the stop of reaction time.

2.4 Analysis of characteristics of liquefied products

2.4.1 Comparison experimental of different acidic catalysts

Experimental liquefaction of waste woody materials were carried out by varying the weight charge ratio of wood to phenol with 1/4 and the reaction temperature at 150°C by oil bath. Into a 500ml three necked flask equipped with stirrer. In addition, the same weight ratio of wood/phenol/ acid catalyst were using in the comparison experimental. Otherwise, the characteristics of produce research experimental was liquefied for 3 hours with two kinds of acidic catalysts, we tested the residues, viscosity, nonvolatility to evaluate the correlation.

2.4.2 Viscosity measurement of liquefied waste woody materials

The viscosity of liquefied woody materials was measured with a viscometer (Model VT-04F, RION Co. Ltd., Japan) in 300ml standard beaker at 25±3°C. The viscosity is measured using a viscosity measurement stick to obtain direct readings in decipascal-seconds (dPa.s).

2.4.3 Measurement of liquefaction residue

The liquefied woody and model substances material were weighted and diluted with methanol, and Whatman filter papers was used to collect the liquefied residue. Here, *Solvent-insoluble residue* in equation (1) means that the weight of liquefied residue through filtration which diluted by methanol from liquefied products, and *raw material* means that the weight of the reactants The liquefied residue was dried in a heating oven at 105°C during 12 hours and calculated the residues by equation (1).

$$\textit{Liquefied rate (wt\%)} = \textit{(Solvent-insoluble residue} / \textit{Raw material)} \times 100\% \quad (1)$$

2.4.4 Measurement of nonvolatility of resultant

About (1.5-3.0)±0.5g of *liquefied products* were weighted and dried using a oven at 180±1°C by 1 hour according to JIS-K6910 to remove the un-reacted phenol. The nonvolatility can be calculated by *residue after oven-dry* according the following equation (2).

$$\textit{Nonvolatility (wt\%)} = \textit{(Residue after oven-dry} / \textit{liquefied products)} \times 100\% \quad (2)$$

2.4.5 Measurement of liquefied products

In this study, we also used the *Liquefied products* to evaluate the liquefaction reaction efficiency and reaction rate. The method of calculation is to subtract the weight of liquefied residence from nonvolatility by the following equation (3)

$$\textit{Liquefied products (wt\%)} = \textit{(Nonvolatility)} - \textit{(Solvent-insoluble residue)} \quad (3)$$

2.5 Molecular weight distribution of liquefied products measured by gel permeation chromatography

The solvent was removed by using a rotary evaporator. The sample was then diluted with 0.1wt% of sample in 99.8% tetrahydrofuran (THF) and filtered through a 0.45μm PVDF filter. The filtrate was analyzed through high performance liquid chromatography (HPLC) using a GPC (HPLC Pump JASCO PU-2080) equipped with RI Detector JASCO RI-2031 and a Shodex GPC KF-806 column which exclusion limit molecular weight (2×10^7). Measurements were conducted at 40°C, 280 nm, and a flow rate of 1 mL/min using 99.8% THF as a mobile phase. The average molecular weight of the samples was calculated by using a calibration curve of monodisperse polystyrene standards. The peak on the GPC curve corresponding to the phenol that was used for the liquefaction was excluded in the calculation of the average molecular weight. As the results, the differential molecular weight distribution (D.MWD) can be used to explain and compare with the analytical results from gel permeation chromatography clearly.

3 Results and discussions

3.1 Characteristics and behavior of liquefied woody material

Kobayashi *et al.* have done the liquefaction of cellulose powder, steamed lignin, alkali lignin, and their mixture was carried out to estimate the liquefaction process of wood in the system using polyethylene glycol as a coexisting medium [12]. They found that Analysis of the mixture showed that its behavior resembled that of wood in changes of molecular weight distribution and the main functional groups. In our study, we analysis the Mw of raw woody powder material can be reach to 1851, and it near to the result of mixture. The model substances of cellulose and lignin, the similar weight distribution was given in results.

3.2 Effect of reaction temperature on liquefaction of model substances

We used model substances mixed by a 5:3 mixtures of cellulose to lignin as simulated coniferous tree and a 6:2 mixture of those as simulated broadleaf tree (figure 1), basing on the composition ratios of the proportion of coniferous and broadleaf trees which according to reference data Nippon Paper Group.

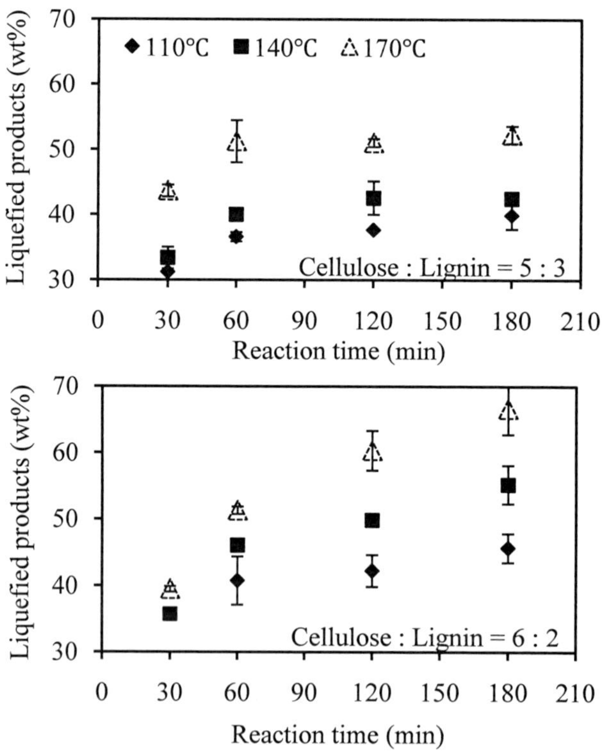

Figure 1: The liquefied products by two kinds of mixture at 3 hour.

According to the results of figure 1, the liquefied products have increased at the two ratios amount of mixture with the rise of reaction temperature. Especially, after the temperature rose to 170°C, the more products pronounced increased at 3 hours. Comparison of two graphs, it can be found that contains a high percentage of the amount of cellulose can get more resultant. In other words, liquefied residual is declining with the reaction temperature. At the first hour, the reaction speed is faster than other time, and it presents a more intense reaction phenomenon.

Zhang *et al.* [8] reported that the temperature of 120°C provides the lowest molecular weight. This temperature, however, does not provide satisfactory liquefaction. As temperature rises, the liquefaction proceeds rapidly and the time reaches to the maximum liquefaction is shortened. At the lower temperature, and the content of cellulose is low, we can get the different molecular weight distribution and lower average molecular weight (figure 2). According to figure 3 and figure 4, the result of molecular weight distribution in present of sulfuric acid catalyst and rate of cellulose and lignin is 5:3 under 3 different temperatures. As the temperature rose, the liquefied products gain of the new condensed residue. These may come from liquefied products of self-condensed polymerization. At the beginning of liquefaction, cellulose degraded to the levulinic acid derivatives, the aromatic derivatives react with them. As a result, the residue was produced. It can be seen from the results, varies with the temperature increased, the product volume also increased. And the products of high molecular weight also increased. Increment of the reaction temperature can promote the condensation of products. These findings clearly indicate that the newly formed residue is originated from the products of cellulose and contains the bound phenol. In other words, phenol plays as a role of self-condensed reagent in the residue, resulting in its weight gain in liquefied products.

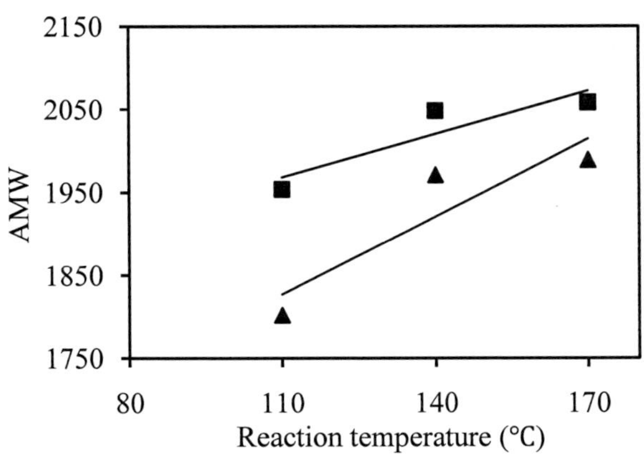

Figure 2: Relationship between average molecular weight (AMW) and reaction temperature. (▲ Cellulose: Lignin = 5:3; ■ Cellulose: Lignin = 6:2).

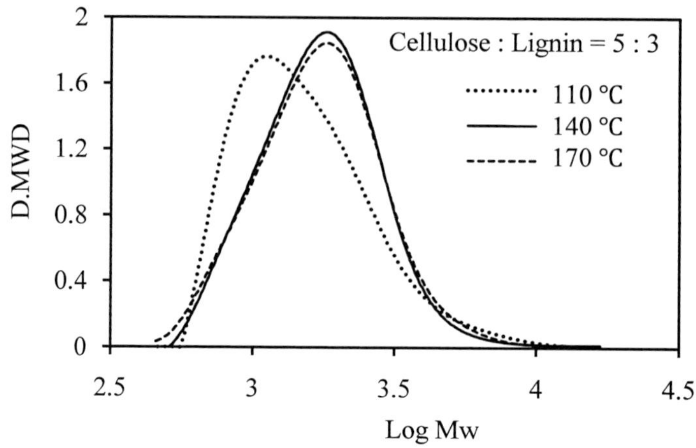

Figure 3: Differential molecular weight distribution (D.MWD) of liquefied products in present of sulfuric acid catalyst with the mixture of cellulose to lignin is 5:3 (simulated coniferous tree) under 110, 140 and 170 °C.

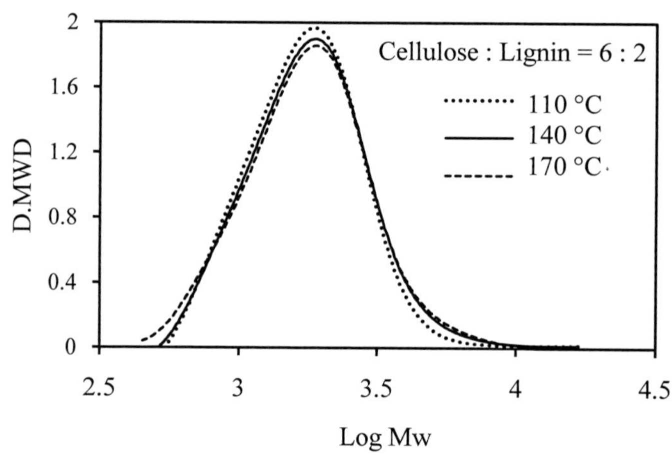

Figure 4: D.MWD in present of sulfuric acid catalyst with the mixture of cellulose to lignin is 6:2 (simulated broadleaf tree).

3.3 Effect of the catalyst on woody liquefaction

In figure 5 and table 3, the result of liquefied products and average molecular weight has been shown. 8% wt and 20% wt of the catalyst amounts were added to different cases respectively. The larger amount of catalyst was added in the two reactions which will lead to react, more liquefied products were produced. However, the molecular weight distribution results showed no change until nearly three hours. It can be considered that the reaction efficiency will be

Table 3: The result of average molecular weight which under the different (8wt% and 20wt% of phenol) amount of sulfuric acid catalyst.

	Cellulose :Lignin=5:3		Cellulose :Lignin=6:2	
	8wt%	20wt%	8wt%	20wt%
Mn	1545	1503	1597	1453
Mw	1971	1942	2048	2150
Mz	2726	2688	2817	3211
Mw/Mn	1.28	1.29	1.28	1.48

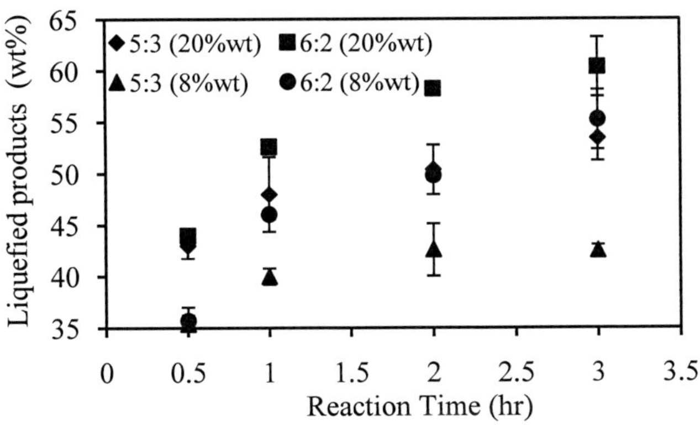

Figure 5: Liquefied products changed under different amounts of catalyst.

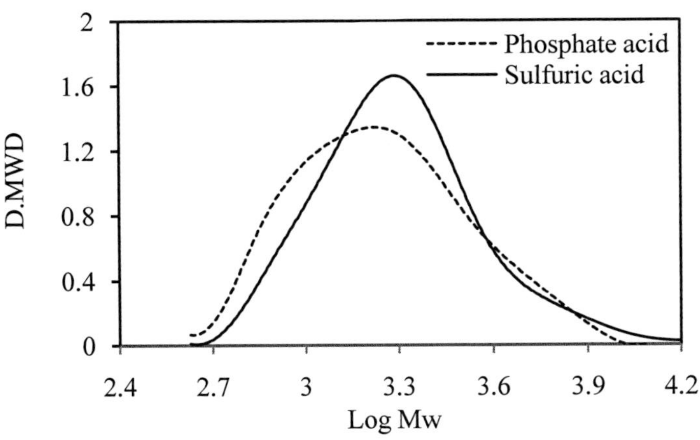

Figure 6: Differential molecular weight distribution of liquefied products in present of sulfuric acid catalyst and phosphate acid catalyst.

increased with the increased of the usage amount of catalyst. Furthermore, the upward tendency of the condensation reaction was not able to be seen. The result of figure 6 having used the different catalyst is compared and in the case of phosphoric acid, molecular weight distribution is wider. The results confirmed that the product affects the chemical structure by different acid catalysts.

3.4 Effect of the reaction time on woody liquefaction

Figure 7 shows the differential molecular weight distribution with reaction time at 30, 60 and 120 min in present of sulfuric acid catalyst. The solid line is the

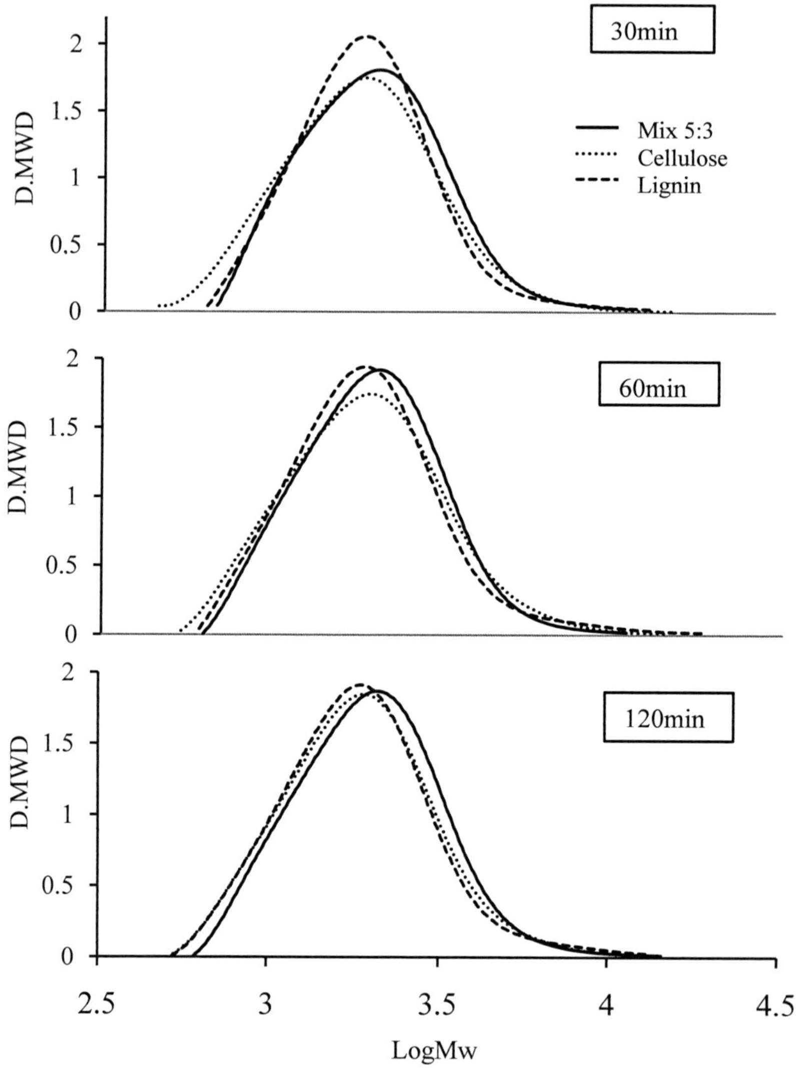

Figure 7: Differential molecular weight distribution of liquefied products in present of sulfuric acid catalyst during 30, 60 and 120 min.

result of mixed case In order to investigate whether it is what a condensation reaction depends on coexistence of the sugar ingredient of cellulose origin, and polyphenol of lignin origin, or it is the characteristic to depend on using woody samples, the last series comparative experiments were conducted.

From the results can be seen on the figure 7, we found that mixed cases, molecular weight distribution curve slightly skewed to the right, and it was reported that the molecular weights of the liquefied products of cellulose were distributed over a wide range of molecular weight, and the derivatives of levulinic acids remained after a long reaction time without condensing. Meanwhile no major functional groups of steamed lignin repolymerized during the liquefaction [12]. In mixed cases, generate new material to improve the molecular weight. By the interaction of the cellulose which carried out solution polymerization, and the aromatic derivative of lignin origin, we thought that a condensation reaction occurred. The upward tendency of the condensation reaction was checked. It is clearly shown that the condensation reaction took place by the mutual interaction between cellulose and lignin. The higher content of cellulose was shown that strengthen condensation reaction. It was considered that detailed mechanism of the condensation reaction should be researched by the development of waste woody polymeric materials [13] for sustainable chemistry.

4 Conclusion

In this study, we examined the relevance of liquefaction conditions, the molecular weight, or molecular weight distribution by using a GPC. It was investigated whether it is what a condensation reaction depends on coexistence of the sugar ingredient of cellulose origin, and polyphenol of lignin origin, or it was a characteristic to depend on using woody samples. When reaction efficiency was raised, the relevance of the output characteristic was analyzed. Within 3 hours of the reaction, it is shown that the temperature will influence on output performance as a change factor of reaction conditions. The relevance of the upward tendency of the amount of catalyst addition and a condensation reaction was not confirmed. The influences of acid catalyst on the structure of liquefied products have been confirmed. When the mixture of cellulose and lignin was liquefied, the condensation reaction occurred, but when the condensation reaction occurred from a part of insoluble residue having been formed by the interaction of the cellulose which carried out solution polymerization, and the aromatic derivative of lignin origin. The upward tendency of the condensation reaction was checked. It is clearly shown that the condensation reaction took place by the mutual interaction between cellulose and lignin. The higher content of cellulose was considered that strengthen condensation reaction. In our future work of woody polymeric materials for sustainable chemistry, we will try to separate the substances from condensation reaction and un-reacted materials.

Acknowledgement

Some works of this study were supported by the special funds for Basic Researches (B) (No. 22404022, FY2010~2012) of Grant-in-Aid for Scientific Research of the Japanese Ministry of Education, Culture, Sports, Science and Technology (MEXT), Japan.

References

[1] T. Asano *et al.*, The effect of the ozonization exerted on liquefaction of the research-wood about resinification and recycling of woody system waste, 2007, Japanese.

[2] L.Z. Lin *et al.*, Liquefaction mechanism of [beta]-O-4 lignin model compound in the presence of phenol under acid catalysis. Part 1. Identification of the reaction products, *Holzforschung*, **55**, 617–624, 2001a.

[3] L.Z. Lin *et al.*, Liquefaction mechanism of [beta]-O-4 lignin model compound in the presence of phenol under acid catalysis. Part 2. reaction behaviour and pathway, *Holzforschung*, **55**, 625–630, 2001b.

[4] L.Z. Lin *et al.*, Liquefaction mechanism of lignin in the presence of phenol at elevated temperature without catalysts. I. structural characterization of the reaction products, *Holzforschung*, **51**, 316–324, 1997a.

[5] L.Z. Lin *et al.*, Liquefaction mechanism of lignin in the presence of phenol at elevated temperature without catalysts. II. Reaction pathway, *Holzforschung*, **51**, 325–332, 1997b.

[6] L.Z. Lin *et al.*, Liquefaction mechanism of cellulose in the presence of phenol under acid catalysis, *Carbohydrate Polymers*, **57**, 123–129, 2004.

[7] M. Kobayashi *et al.*, Analysis on residue formation during wood liquefaction with polyhydric alcohol, *J Wood Sci*, **50**:407–414, 2004.

[8] Y.C. Zhang *et al.*, Characterization of liquefied product from cellulose with phenol in the presence of sulfuric acid, *Bioresource Technology*, **97**, 313–321, 2006.

[9] Sung P.M *et al.*, 2006, Effect of orangic sulfonic acids as catalysts during phenol liquefaction pinus radiate bark, *Ind Eng. Chem*, **12**, 720–726.

[10] T. Yamada *et al.*, 2001, Characterization of the products resulting from ethylene glycol liquefaction of cellulose, *J Wood Sci*, **47**, (6), 458–464.

[11] Sung P.M *et al.*, 2009, Liquefaction of cellulose in the presence of phenol using p-toluene sulfonic acid as a catalyst, *Journal of Industrial and Engineering Chemistry,* **15**, 743–747.

[12] M. Kobayashi *et al.*, 2004, Analysis on residue formation during wood liquefaction with polyhydric alcohol, *The Japan Wood Research Society,* **50**, 407–414.

[13] Q. Wang *et al.*, 2010, Liquefaction processes and characterization of liquefied products from waste woody materials in different acidic catalysts, *The Sustainable World, WIT Transactions on Ecology and the Environment*, **142**, 343-354.

Basic study on combustion characteristics of waste rice husk and emission behavior from a new-type air vortex current combustor

Q. Wang[1], T. Maezono[1], Q. Chen[1], P. Apaer[1], Y. Wang[1], L. Gui[1], D. Niida[1], N. Mitsumura[1], M. Domon[2], I. Fujiwara[2] & N. Yamaguchi[2]
[1]*Department of Environmental Science and Technology, Graduate School of Science and Engineering, Saitama University, Japan*
[2]*Kaneko Agricultural Machinery Co, Ltd., Japan*

Abstract

There are large quantities of rice husk estimated around 3 million tons as agricultural waste every year in Japan. Air pollutants emitted from exhaust gases of rice husk incineration lead to very important environmental damage, not only because of the influence on global environment and climate, when released into the atmosphere, but also on human health due to the local air pollution. Therefore, it is necessary to effectively utilize the rice husk waste and to reduce the air pollution. We try to develop a new-type air vortex current small-scale combustor which can effectively combust rice husk as biomass energy instead of fossil oil fuel for farming-greenhouses heating during the winter season. In this study, we investigated if rice husk can be fed on the new-type air vortex current small-scale combustor and reduced fossil fuel. The new-type small-scale combustor is able to keep a constant high temperature (about 1000°C) even if the rice husk combustion is not under the best conditions.

At the same time, it is also important to evaluate the emission behavior of harmful air pollutants emitted from the rice husk combustion with measuring carbonaceous and ionic composition of suspended particulate matter (SPM) in the exhaust gases from the new-type air vortex current combustor, and to reduce the pollutant emission by controlling the combustion conditions. From the analytical results of the size distribution of carbonaceous composition collected by an air sampler, it is shown that elemental carbon dominated in the coarse

WIT Transactions on Ecology and the Environment, Vol 154, © 2011 WIT Press
www.witpress.com, ISSN 1743-3541 (on-line)
doi:10.2495/CHEM110191

particles, which are produced by incomplete combustion, and organic carbon dominated in the fine particles. Carbonaceous concentrations can be reduced substantially in the emitted particles by highly effective combustion when the combustor was improved. As the results of the ionic composition, high concentrations of potassium ion as a tracer of biomass burning were determined. Combustion temperature control is important to avoid corrosion in the system and the health effects from high concentrations of chloride contents.

Although the new-type air vortex current combustor developed in our purpose is typically small-scale, however, usual fixed-bed combustors are prone to be incomplete because of the simplicity of the structure. Since there are no specific regulations for these kinds of combustors in Japan, therefore, even small combustor fall out of the possibly applicable emissions regulations, to ensure stable combustion performance and less air pollutants. In near future, we try to improve the combustor fed with less fossil oil fuel and more rice husk waste which will be feasible and sustainable.
Keywords: rice husk, combustion characteristics, small-scale combustor, air vortex current, exhaust gases, suspended particulate matter (SPM), carbonaceous and ionic analysis.

1 Introduction

Currently, global warming is become increasingly evident in the global climate. Combustion of fossil fuel is generally admitted as the main responsible for global warming. Though, the use of fossil fuel is expected to increase in the future because of economic development and growth of population in developing countries [1], hence, the only solution is zero-emission technology, that is, to reduce and minimalize all possible emissions produced by human activities [2]. In order to achieve zero-emissions, it is important to apply a technology to utilize all unused waste biomass [3, 4].

In Japan, waste biomass and residues produced from agricultural and forestry activities have been unused and mostly being incinerated for disposal, due to its high cost of collection, transport, and storage and also the needs of energy that it implies. Moreover, urgent measures are required to reduce the effects to air pollution from illegal incineration.

It is estimated that around 3 million tons of waste rice husk which is the most common agricultural residue are wasted every year in Japan. Additionally, since rice is the staple food and regular part of the diet for almost half of the world population, an effective utilization of waste rice husk will be an important countermeasure to global warming. For these reasons, in the recent years, there is an increasing demand on the utilization of unused biomass instead of usual fossil oil fuel combustors for farming-greenhouses heating during the winter season. This increase in the demand will make prices to increase. In general, these combustors are small-scale [5], therefore, the established regulations (e.g. Japanese air pollution control act and waste disposal and public sanitation regulation) cannot apply for this kind of pollution emission control. So far, small-scale combustors are characterized by simplicity on their structure and low cost and hence emit visible black smoke due to their low combustion performance because of the lack of the regulations [6, 7].

In this study, we investigated if fossil fuel can be substituted by a new-type air vortex current small-scale combustor of the waste rice husk. This new-type combustor is shown in figure 1. In order to use the waste rice husk samples as the fuel in the new-type air vortex current combustor, we analyzed the chemical composition of the waste rice husk as agricultural waste and investigated its combustion characteristics. Also harmful substances emitted from the rice husk combustion were evaluated by measuring the suspended particulate matter composition on the exhaust, and the reduction possibility of these harmful substances by controlling the combustion conditions was examined.

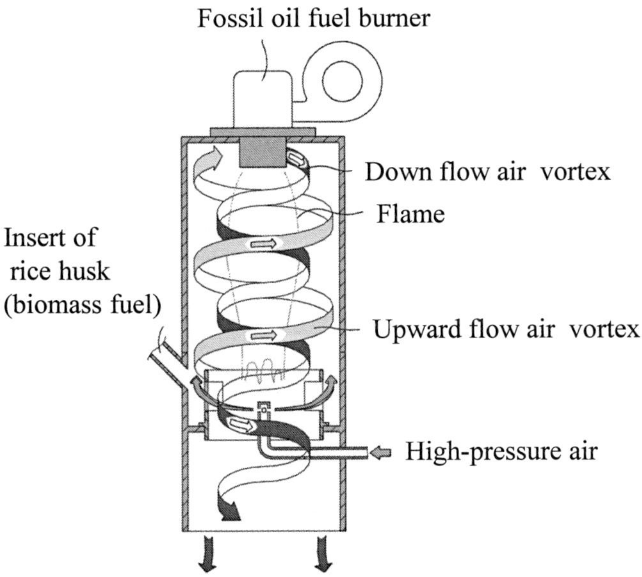

Figure 1: Concept of the new-type air vortex current small-scale combustor.

2 Experimental methods

2.1 Composition analysis of rice husk samples as agricultural waste

In this study, six different rice husk varieties were analyzed. The proximate and ultimate analyses of rice husk samples from different producing areas (brands and farms) were carried out according to the Japanese industrial standard (JIS) method of JIS-M8812.

2.2 Evaluation method for combustion of the waste rice husk samples

Combustion characteristics of rice husk samples from different producing areas were analyzed by the thermogravimetric/differential thermal (pyrolysis) analysis

(TG/DTA), and under the following conditions: about 1.0 mg of sample was heated at a rate of 5 °C min^{-1} starting from room temperature until 600 °C. The gas flow rate was 250 mL min^{-1}, clean air was used as carrier gas for combustion.

2.3 Combustion control method

A high temperature resistant cylindrical small-scale combustor with a diameter of 20 cm was set up. A fossil oil fuel burner was set in the upper side while the waste rice husk was fed in the lower part. The combustion was started first with oil fuel alone and the rice husk was supplied when the reactor wall reached a temperature over 480 °C. Combustion temperature was monitored with a thermocouple and controlled by the supplied air flow rate. The fossil oil fuel supply was stopped when the small-scale combustor wall reached a temperature around 550 °C and combustion was maintained only by the supply of the waste rice husk samples.

2.4 Evaluation of suspended particulate matter (SPM) in exhaust gases

2.4.1 Air sampling method for exhaust gases collection

The air sampling system of exhaust gases is shown in figure 2. In order to evaluate the suspended particulate matter (SPM) emitted from the combustor, SPM were collected on quartz-fiber filters (80 mm diameter, pallflex prpducts corp, 2500QAT-UP) and Teflon filters (8 mm diameter, pallflex prpducts corp, TX40HI20-WW) using two air samplers which are called the low pressure impactor (LPI: Model LP-20, Tokyo Dylec Corporation).

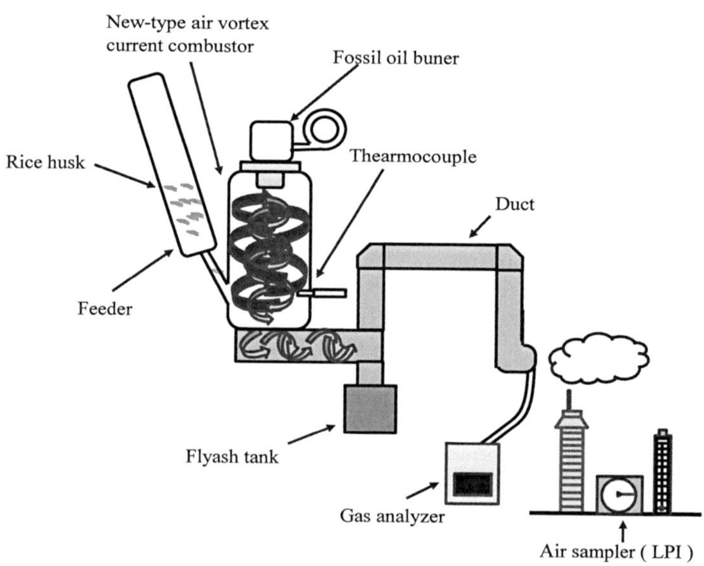

Figure 2: Air sampling method for exhaust gases emitted from the combustor.

Suspended particulate matter (SPM, separated particle sizes< 0.06, 0.06-0.12, 0.12-0.20, 0.20-0.30, 0.30-0.50, 0.50-0.70, 0.70-1.20, 1.2-2.1 and 2.1-3.5 μm) were collected respectively at a flow rate of 23.6 L min^{-1} for 20 min on each sampling with LPI air samplers. The quartz-fiber filter samples were used for carbonaceous composition analysis and samples, and the Teflon filters were used for ionic composition analysis. Size distribution, characteristics and chemical components of SPM were measured because of the influence on global environment and climate, when released into the atmosphere, but also on human health due to the local air pollution.

2.4.2 Evaluation of carbonaceous composition of SPM in the exhaust gases

In generally, coarse particles of suspended particulate matter (SPM) are mainly having a diameter over 2 μm and it unable to instruction in to entering the respiratory tract by the nose, throat, and pharynges. However, fine particles with a diameter under 2 μm are able to reach deeper parts of the respiratory system, including the air sacs of the lungs. While particles with a diameter below 0.1 μm (or ultra fine particles) are able to break into the alveoli and through the capillary beds reach the blood stream. That suspended particulate matter is formed by carbonaceous material. Therefore it is necessary to analyze the size distribution of carbonaceous composition emitted from the combustion of rice husk. Carbonaceous analysis for elemental carbon (EC) and organic carbon (OC) was based on the IMPROVE method (Interagency Monitoring of Protected Visual Environment) by the thermo-optical carbon analyzer (Model 2001, Thermo/Optical Carbon Analyzer, Desert Research Institute) shown in table 1.

In this method, a 0.503 cm^2 (8 mm diameter) punch aliquot of each air sampling quartz filter was heated at 120°C (OC1), 250°C (OC2), 450°C (OC3), and 550°C (OC4) in a helium (He) atmosphere, and then at 550°C (EC1), 700°C

Table 1: Protocol of IMPROVE thermal / optical method for carbonaceous analysis.

Carbonaceous fraction	Temperature (℃)	Atmosphere
OC1	120	100% He
OC2	250	
OC3	450	
OC4	550	
EC1	550	$2\%\ O_2 + 98\%$ He
EC2	700	
EC3	800	

(EC2), and 800°C (EC3) in an oxidizing atmosphere of 2% O_2 and 98% He. The analysis was repeated two or three times for each air sampling quartz filter for better accuracy.

2.4.3 Evaluation of ionic composition of SPM in the exhaust gases

One fourth of 8 mm diameter Teflon filter was ultrasonically extracted with 50 mL ultrapure water (18.2MΩ milli-Q water) for 20 minutes, in order to carry the ionic composition analysis of anions (NO_3^-, SO_4^{2-}, and Cl^-) and cations (K^+, Na^+, NH_4^+, and Ca^{2+}), respectively. All extracts were analyzed using an ion chromatograph (Dionex-100 IC, Japan Dionex Co., Tokyo, Japan) equipped with an electric conductivity detector. Mixed standards of five concentrations within the ranges of 0.1-2.5 ppm (w/v) for anions and 0.05-0.8 ppm (w/v) for cations were used to draw standard calibration curves. The analysis was also repeated two times for each air sampling quartz filter for better accuracy.

3 Results and discussions

3.1 Measurements in the composition of the rice husk

In table 2 are shown the composition analysis for the waste rice husk samples from different producing areas. Regarding the residual ash contents of the waste rice husk slightly below 20% in mass, it was supposed that the waste rice husk is one kind of low-grade fuels, and it should be difficultly to be widely used as fuel [8].

Table 2: Composition analysis of rice husk samples from different producing areas (names of farm or brand).

Producing areas Farm (brand)	Ash (wt%)	M (wt%)	VM (wt%)	FC (wt%)	C (%)	H (%)	N (%)
Miyazaki (Koshihikari)	14.3	9.9	61.3	14.6	38.7	5.1	0.4
Kirari Miyazaki	13.3	10.0	61.9	14.8	39.0	5.1	0.3
Niigata	18.9	8.4	58.2	14.5	37.4	4.7	0.3
Hokkaido	16.7	9.7	60.9	12.7	39.2	5.1	0.5
Saitama	18.4	6.70	60.0	14.9	38.8	5.1	0.4
Hokkaido (Hoshinoyume)	19.2	7.6	59.2	14.0	37.9	5.1	0.5

M: Moisture, VM: Volatile matter and FC: Fixed carbon.

3.2 Differential combustion characteristics of the different rice husk samples

The thermogram for Miyazaki rice husk (Koshihikari) showed two well-defined peaks of TG/DTA at 320°C and around 400°C (figure 3). These results show that this kind of waste rice husk achieves its pyrolysis around 320 °C, where the more volatile components are burned and the carbonized fraction is burned at a higher temperature, around 400°C. This combustion behavior was also confirmed for the other three kinds of rice husk with the peak differences around 20 to 50°C between the different waste rice husk samples produced from various areas (farms or brand).

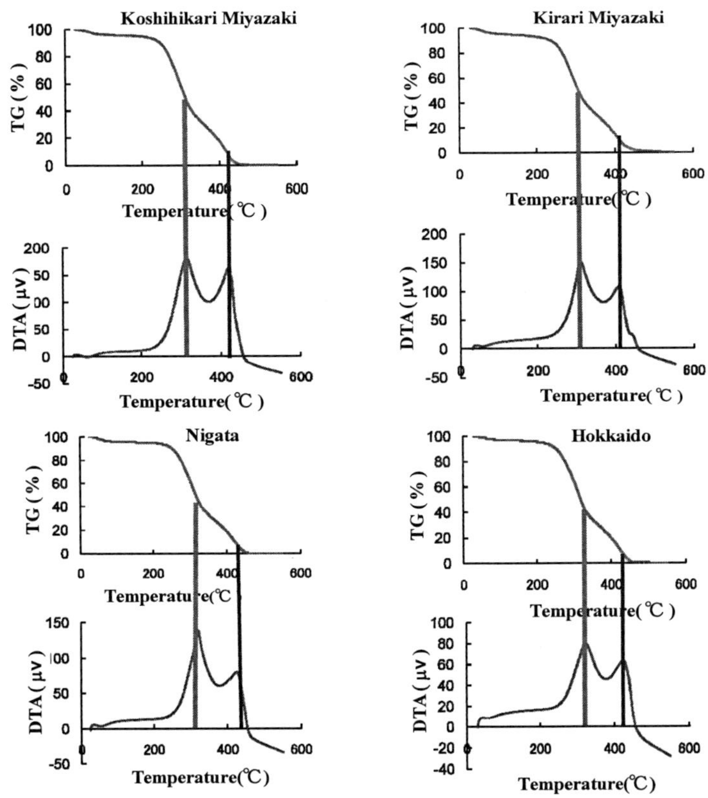

Figure 3: Combustion and pyrolysis behavior of rice husk samples.

3.3 Temperature control for the new-type combustor of waste rice husk

In order to reach stabilization in the combustor temperature, the combustor was first heated with the oil burner until the temperature of 550 °C was reached, then the rice husk was fed into the combustor which made the temperature to change. After that, it is waited for the stabilized temperature until about 800 °C in the outer wall and/or about 1,000 °C in the inside of the combustor. Figure 4 shows the variation of the temperature profiles in the outer wall and the inside of combustor and surrounding air in the room. Temperature in the inside of the combustor is stabilized around 970 °C, which is the well-known temperature condition that the harmful substances such as dioxins can be extremely reduced.

In addition, if temperature is controlled at those levels, the generation of thermal-NOx can also be controlled because thermal-NOx is generated at temperatures above 1,100°C. The combustion temperature and the heating energy provided by this new-type combustor is enough for the intended purpose, additionally, it can be achieved with trace levels of harmful substances and less NOx emissions.

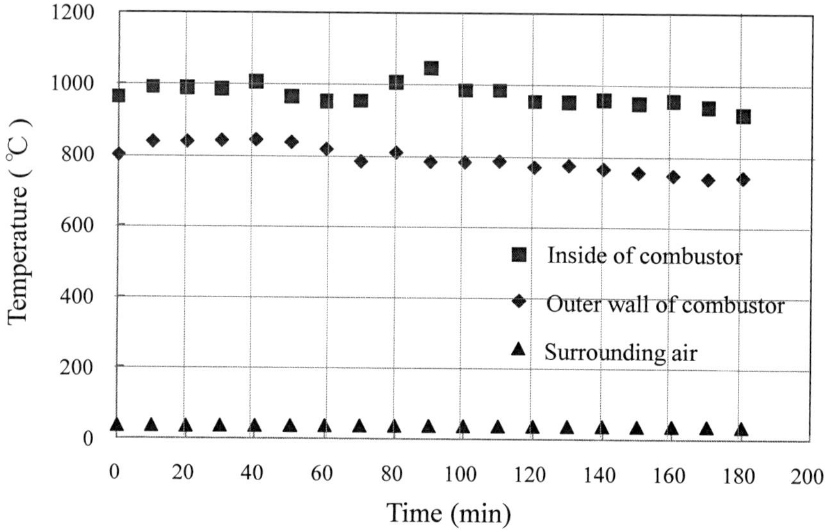

Figure 4: Temperature profiles in the inside and outer wall of the combustor and surrounding air during analysis.

3.4 Evaluation of exhaust gases from the new small scale combustor

3.4.1 Emission behavior of carbonaceous composition of SPM in the exhaust gases

The analytical results of carbonaceous composition of suspended particulate matter (SPM) in the exhaust gases are shown in figure 5. EC1 which is mainly generated by biomass burning (combustion) at relatively lower temperatures was found at the highest concentrations in the particle sizes from 1.2-2.1 µm to 2.1-3.5 µm as the coarse particles. OC4 was also determined at the high levels in fine particles (1.2-2.1 µm) and coarse particles. From these results, we can conclude that these combustion conditions of the waste rice husk may also be under incomplete combustion. OC2 includes compounds like levoglucosan and methoxyphenol, which are generated in the pyrolysis of cellulose and lignin, was found at the highest concentrations in the particle sizes below 1.2 µm as the fine particles; levoglucosan is one of the water-soluble organic carbonaceous components and it can contribute to cloud condensation nuclei, and influence on the global warming and climate by the optical properties of such kind of carbonaceous aerosol [9, 10].

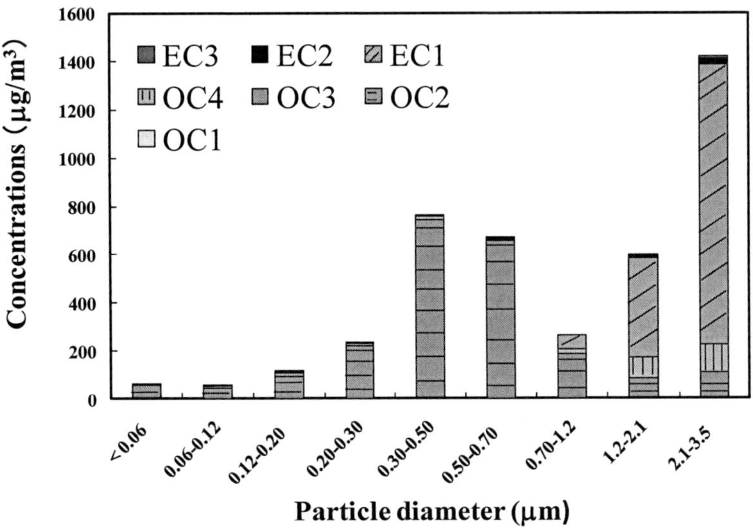

Figure 5: Size distribution of organic and elemental carbon composition in exhaust gases.

We consider that in this experiment, the air sampling method for atmospheric aerosol is not able to measure those semi-volatile carbonaceous components which would condensate from gases to particles by partition under cooling down to ambient temperature, and hence the real contribution to atmospheric aerosol may be underestimated. It is necessary to carry out experiments with an improved the air sampling method for the exhaust gases after air dilution in order to cool the exhaust gases down and have a better measurement of the real contribution of the semi-volatile carbonaceous components in the exhaust gases to atmospheric aerosol.

3.4.2 Emission behavior of ionic composition of SPM in the exhaust gases

The analytical results of ionic composition of suspended particulate matter (SPM) in the exhaust gases are shown in figure 6. Ionic concentration results showed that K^+ had a high concentration in all particle sizes. K^+ is an important component of biomass burning [11], since it is used in metabolic growth processes, hence this component is used as a marker for biomass burning (combustion) contribution to atmospheric aerosol. In addition, Cl^- also showed high concentration in the all particle sizes from combustion of the waste rice husk. Because Cl^- is especially found at higher concentrations during winter seasons; therefore, it is thought to arise from biomass burning of agricultural wastes including waste rice husk. In our further study, we plan to carry out combustion experiments using different biomass fuels, apart from waste rice husk in order to compare the abundance of Cl^- from biomass burning of different agricultural wastes.

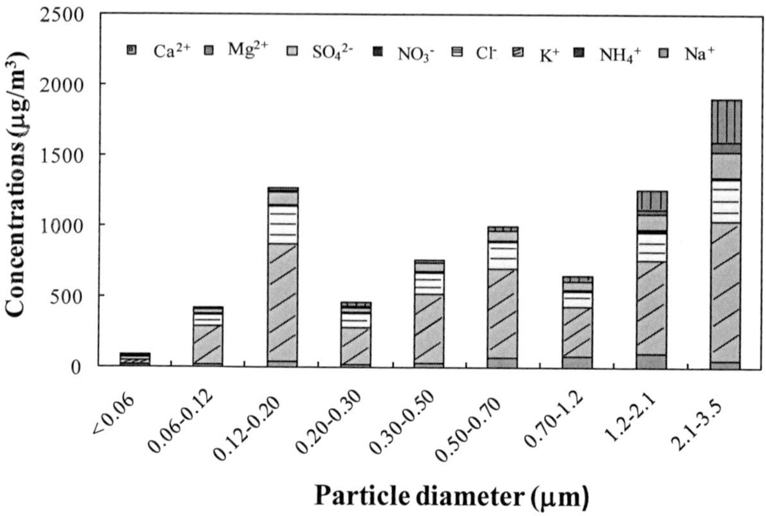

Figure 6: Size distribution of ionic components in exhaust gases.

4 Conclusion

In our experiments, we evaluated the combustion characteristics of the size distribution of exhaust particles (SPM) when using the several kinds of waste rice husk samples from different producing areas as the biomass fuel used for the new-type air vortex current combustor developed by our laboratories. The waste rice husk was suitable to maintain a combustion temperature above 800 °C. Our results indicate that waste rice husk is able to substitute fossil fuels if we use this new-type combustor.

From the analytical results of the size distribution of carbonaceous composition collected by the air sampling system, it is shown that elemental carbon (EC1) dominated in the coarse particles, which are produced by incomplete combustion, and organic carbon (OC2) dominated in the fine particles. Carbonaceous concentrations in the emitted particles can be reduced substantially by highly effective combustion when the combustion conditions of the combustor were improved. As the results of the ionic composition, high concentrations of potassium ion (a tracer of biomass burning) were determined. Combustion temperature control is important to avoid corrosion in the system and the health effects from high concentrations of chloride compounds (ions).

We also have to calculate the better air/fuel ratio in order to reduce the emission of harmful substances [12, 13] and to optimize the fuel/carbon discharge coefficient in order to make the waste rice husk as a renewable energy source for alternate fossil oil fuel in the future.

Acknowledgements

This study was supported by the Special Funds of the local Government of Saitama Prefecture, Japan, and partly subsidized by two Funds for Basic Researches (B) (No. 19404021, FY2007~FY2009 and No. 22404022, FY2010~2012) of Grant-in-Aid for Scientific Research of the Japanese Ministry of Education, Culture, Sports, Science and Technology (MEXT), Japan.

References

[1] A.A. Khan, W.de Jong, P.J. Jansens, H. Spliethoff., Biomass combustion in fluidized bed boilers: Potential problems and remedies., *Fuel Processing Technology*, **90**, 21-50, 2009.

[2] Li Dong, Shiqiu Gao, Guangwen Xu., No reduction over Biomass Char in the Combustion Process., *Energy Fuels*, **24**, 446-450, 2010.

[3] Lone A. Hansen, Hanne, P. Nielsen, Flemming J. Frandsen, Kim Dam-Johansen, Steffen Horlyck, Asger Karlsson., Influence of deposit formation on corrosion at a straw-fired boiler., *Fuel Processing Technology*, **64**, 189-209, 2000.

[4] E.F. Iliopoulou, E.V. Antonakou, S.A. Karakoulia, I.A. Vasalos, A.A. Lappas, K.S. Triantafyllidis., Catalytic conversion of biomass pyrolysis products by mesoporous materials: Effect of steam stability and acidity of Al-MCM-41 catalysts, *Journal of Chemical Engineering*, **134**, 51-57, 2007.

[5] L. S. Johansson, C. Tullin, B. Leckner, P. Sjovall., Particle emissions from biomass combustion in small combustors, *Biomass and Bioenergy*, **25**, 435-446, 2003.

[6] Bernd R.T. Simoneit., Biomass burning-a review of organic tracers for smoke from incomplete combustion, *Applied Geochemistry*, **17**, 129-162, 2002.

[7] Henrik Wiinikka, Rikard Gebart., Experimental investigation of the particle emissions from a small-scale pellets combustor, *Biomass and Bioenergy*, **27**, 645-652, 2004.

[8] W. Permchart, V.I. Kouprianov., Emission performance and combustion efficiency of a conical fluidized-bed combustor firing various biomass fuels, *Bioresource Technology*, **92**, 83-91, 2004.

[9] Jorger. Jimenez, Candiss. Claiborn, Ranils. Dhammapala, Christopherd. Simpson., Methoxyphenols and Levoglucosan Ratio in PM2.5 from Wheat and Kentucky Bluegrass Stubble Burning in Eastern Washington and Northern Idaho, *Environment Science and Technology*, **41**, 7824-7829, 2007.

[10] Zhang Yuan-xun, Shao Min, Zhang Yuan-hangl, Zeng Li-min, He Ling-yan, Zyu-Bin, Wei Yong-jie, Zhu Xian-lei., Source profiles of particulate organic matters emitted from cereal straw burnings, *Journal of Environmental Sciences*, **19**, 167-175, 2007.

[11] J. M. Jones, L. I. Darvell, T. G. Bridgeman, M. Pourkashanian, A. Williams., An investigation of the thermal and catalytic behavior of

potassium in biomass combustion., *Proceedings of the Combustion Institute*, **31**, 1955-1963, 2007.

[12] Q. Wang, P. Aparu, S. Nakamura, T. Maezono, I. Fujiwara and M. Domon, Evaluation of hybrid combustor of biomass and oil by vortex current of air, *Research Report of Comprehensive Open Innovation Center of Saitama University, Japan*, **2**, 105-107, 2009 (in Japanese).

[13] I. Fujiwara and M. Domon, K. Morita, N. Yamaguchi, Q. Wang, T. Kumagai, Evaluation of hybrid combustor of biomass and oil by vortex current of air, *Research Report of Saitama Industrial Technology Center, Japan*, **8**, 1-7, 2010 (in Japanese).

Section 6
Environmental health issues

Possibilities and limitations of LCA for the evaluation of soil remediation and cleanup

V. Cappuyns
Hogeschool-Universiteit Brussel, Centre for Economics and Corporate Sustainability (CEDON), Belgium

Abstract

When evaluating remediation technologies for contaminated soil and groundwater, the beneficial effect of the remediation, namely cleaner soil and groundwater, are mostly emphasized without consideration of the environmental impact of the remediation activities themselves. Nevertheless, practitioners and decision makers can rely on a broad range of decision tools that can help them to achieve a better balance between economic, social and environmental health aspects of contaminated land remediation. A holistic approach for the management of contaminated land should ideally include an assessment of the environmental risk of the contamination, an assessment of the environmental, social and health impact of the remediation process and a cost-benefit analysis of the remediation project.

A life cycle framework, including a life cycle management (LCM) approach structuring environmental activities and life cycle analysis (LCA) for a quantitative examination, can be helpful for the selection of site remediation options with minimum impact on the ecosystem and human health. During the last 10 years, several instances have emerged in which a life cycle approach has been applied to the remediation of contaminated sites. Besides addressing the environmental impact of the remediation activities for a specific site, attention should also be paid to the consequence of reintroducing a remediate site into the economy. From a legal point of view, there should be ways to encourage the use of sustainable remediation technologies, together with a disconnection of treated soils from the definition of waste. Finally, the focus should move from remediation, whether sustainable or not, to the prevention of soil contamination..
Keywords: environmental risk, heavy metals, impact assessment, life cycle management, soil contamination, sustainability.

www.witpress.com, ISSN 1743-3541 (on-line)
doi:10.2495/CHEM110201

1 Introduction

1.1 Policy on soil remediation in Europe

Approximately 250,000 sites in Europe require cleanup, while the European Environmental Agency estimates that nearly 3 million sites are potentially polluted [1]. Industrial activities are responsible for over 60% of Europe's soil pollution (the oil sector accounts for 14% of this total). Among the most common harmful contaminants are heavy metals (37%) and mineral oils (33%) [1]. Although several EU directives support the prevention and cleanup of soil contamination (e.g. EU Directive on Environmental Liability, EU Waste Framework Directive, EU Water Framework Directive, EU Integrated Pollution Prevention and Control Directive), there is no general European directive with regard to Soil Remediation and cleanup. Because the cleanup of all historically-contaminated sites to background concentrations or levels suitable for all types of land use is not considered technically or economically feasible, cleanup strategies are more and more designed to use sustainable, long-term solutions, often using a risk-based approach to land management.

In some countries the selection of soil remediation options is based on the best available technology (BAT). Some other countries, including Flanders (Belgium), take the best available technology not exceeding excessive cost (BATNEEC) as a criterion for selecting soil remediation options. Despite the fact that BAT is likely to represent the most technologically feasible tool for soil remediation, BATNEEC seems a more suitable framework for decisions regarding soil remediation projects.

Sorvari et al. [2] point to the fact that existing policy instruments do not promote sufficiently eco-efficient site remediation technologies. Besides the development of a more eco-efficient policy for contaminated site management, there is a need to promote the use of more eco-efficient site remediation techniques. An important aspect of this is the development of methods that allow assessment of the eco-efficiency of site remediation options. Additionally, the knowledge of decision support tools that take into account eco-efficient or even sustainable remediation options, among stakeholders that have to decide on which site remediation options to pursue, should be improved.

1.2 Selection of remediation techniques for contaminated land

As stated above, the decision process for selecting the technique for remediation of contaminated sites has traditionally focused on the cleanup level, the time required for the remediation, economic resources and the best available technologies. Other factors that are relevant for the remediation of contaminated sites include the sustainability of the remediation process and the view of the stakeholders on the remediation solution [3]. Sustainability of the remediation of contaminated land takes into account 3 main aspects: (1) economic aspects (e.g. cost of remediation); (2) ecological aspects (e.g. environmental aspects of remediation such as emissions to surface waters) and (3) social aspects (e.g. communication to stakeholders). Although soil remediation is often considered to be a completely positive process because of the reduction or removal of soil contamination, the overall consequences and impact of the soil remediation process should also be considered. Under the European Commission's Framework Programme 5, the Concerted Action CLARINET brought experts and regulators from 16 European countries together to identify means by which contaminated land could be managed in an effective but sustainable manner. As a follow-up action to implement the recommendations of this Concerted Action, the SNOWMAN project is to analyze the possibilities for improving co-operation in Europe between national research programs in the field of soil and groundwater management where there is the presence of contamination [4]. The European Coordination Action for Demonstration of Efficient Soil and Groundwater Remediation (EURODEMO) promotes sustainable, cost-effective soil and groundwater remediation technologies. This leaves us with the fundamental question: what constitutes sustainable remediation of contaminated soil?

Harbottle et al. [5] defined 5 criteria that have to be met in order to classify a remediation project as 'sustainable'(Table 1).

Table 1: Five criteria for a sustainable remediation project (after [5]).

	Criterion	*Comment*
1	Future benefits outweigh cost of remediation	Besides the economic cost, societal benefits also have to be taken into account.
2	Overall environmental impact of the remediation process is less than leaving the land untreated	Boundaries of the impact assessment have to be defined: which impacts do you take into account and which not?
3	Environmental impact due to the application of the remediation process is minimal and measurable	Boundaries of the impact assessment have to be defined. There should be a consensus on the way to express 'environmental impact' in an objective way.
4	The time-scale over which the environmental consequences occur (including intergenerational risk) is part of the decision making process.	The soil should be restored in a sustainable manner so that both current and future generations can satisfy their ecological, economic and social needs
5	The decision-making process includes an appropriate level of engagement of all stakeholders	What can the role of different stakeholders be in every aspect of the decision making process

Sustainable soil remediation can also be addressed from the point of view of the sustainable use of resources. Both soil and groundwater can be considered valuable resources. Ideally, remediation and/or cleaning of soil and groundwater should be performed in a closed-loop system, with conservation of landscape characteristics, to minimize the environmental impact of the remediation project and to achieve the goal of 'sustainable use of soil'.

The term "green or gentle remediation techniques" is closely related to "sustainable remediation, as "green remediation techniques" are defined as remediation techniques with a lower environmental impact and a lower associated consumption of natural resources such as water and energy [6]. Nevertheless, only one aspect of "sustainability", namely, the environmental aspect, is taken into account.

In the present paper, attention is first paid to the assessment of the environmental impact of the soil remediation process and on the way this environmental impact is quantified by means of life cycle assessment (LCA) methodology. On the one hand, some possibilities of using LCA for the evaluation of soil remediation are illustrated with examples from literature. On the other hand, some additional aspects that should be accounted for in the sustainable remediation of contaminated land and that are not or only partly addressed in LCA, are discussed.

2 Evaluation of the environmental impact of soil remediation by LCA

The environmental impact of soil remediation is one of the aspects that deserve more attention during soil remediation. Since the 1990's, several tools have been developed to assess the environmental impact of processes and products, such as eco-indicators and other tools based on Life Cycle Assessment (LCA). Since the last decade, LCA has been gaining wider acceptance as a tool for the quantification of environmental impacts and evaluation of improvement options throughout the life cycle of a process, product or activity [7]. A Life Cycle Analysis is carried out in four distinct phases: (1) definition of goal and scope, (2) Life cycle inventory, (3) life cycle impact assessment and (4) interpretation [8]. Historically, LCA has mainly been applied to products, but several examples (e.g. [9]) show that it can assist in identifying more sustainable options in process selection design and optimization.

A life cycle framework, including a life cycle management (LCM) approach structuring environmental activities and life cycle analysis (LCA) for a quantitative examination, can be helpful for the selection of site remediation options with minimum impact on the ecosystem and human health [10].

2.1 Application of LCA to the remediation of contaminated sites

During the last 10 years, several instances have emerged in which a life cycle approach has been applied to the remediation of contaminated sites.

In a literature review, Suér et al. [11] illustrate that the result of LCA is highly dependent on the method used and that the choice of impact categories heavily affects the outcome of an LCA study. An updated overview (Table 2) of case studies dealing with LCA of the remediation of contaminated sites illustrates the high variation in impact assessment methods that have been used over the last

Table 2: Literature overview of LCA-based case studies where the impact of soil remediation technologies were evaluated.

Ref	Case study	Impact assessment method or tool
[19]	-	REC (uses value functions method for assessment of environmental merit)
[20]	-	Calculation of potential impact indicators
[10]		Multimedia Maclkay model Solid Waste Burden (SWB) + useable land area
[17]	Site contaminated with mineral oil, PAH and Cr	Use of 'disadvantage factors
[21]	-	pollution factor (PF) is calculated, and expression of environmental impacts in dimensionless environmental impact units (EIUs
[18]	Industrial site contaminated with sulfur	No impact assessment but ranking of productivity resources
[22]	spent pot lining (SPL) landfill contaminated with Cd and Cu	EDIP97 + simulation of contaminant transport in groundwater, using site-specific data
[12]	Diesel-contaminated site	EDIP97
[23]	Old landfill	Specific method for impact assessment transport of heavy metals
[24]	former manufactured gas plant site	Characterisation method adopted from UBA (2000),
[5]	Mixed industrial-residential-commercial area with BTEX and THP contamination	IPPC Tier Two methodology
[25]	Brownfield contaminated by human activity in railway sector	IMPACT2002+
[14]	Diesel-contaminated site	US-EPA TRACI
[26]	Industrial site with 300 industries involved in chemical and petro-chemical productions	DEcision Support sYstem for REhabilitation of contaminated sites (DESYRE).
[16]	Previous oil depot	ReCiPe EPD
[15]	Industrial site with distribution centre for new cars	REC

10 years. Besides primary impacts, associated with the state of the site and secondary impacts, associated with the site remediation itself [12], LCA for contaminated site management should also account for tertiary impacts, associated with the effects of the reoccupation of the site [13]. Therefore,

different scenarios could be considered and the collection of additional data concerning temporal and spatial effects should be integrated into the evaluation of contaminated sites [11]. For example, Cadotte et al. [14] considered two different remediation scenarios: one based on a fast treatment time and another one based on a low environmental impact.

Cappuyns et al. [15] compared two remediation scenarios for a site with a LNAPL contamination by applying the BATNEEC (Best available technique not entailing excessive costs) method and the LCA-based REC (Risk Reduction, Environmental Merit and Costs) method. They concluded that, although an LCA-based evaluation method is much more complex and requires much more data than a classical BATNEEC analysis, both evaluation tools could be used in a complementary way. A preliminary selection of remediation technologies could be based on a BATNEEC analysis, followed by a detailed analysis of the selected remediation options by means of LCA. When alternatives for soil remediation are compared, one should be aware that environmental effects occur on very different environmental problems and geographical scales [16], pointing to the importance of including land use in LCA.

It also clear from Table 2 that most case studies deal with sites contaminated with organic contaminants. Sites contaminated with heavy metals [17, 22] or sulfur [18] are only the subject of a few case studies.

3 Other aspects not or only partly addressed in LCA

Practitioners and decision makers can rely on a broad range of decision tools that can help them to achieve a better balance between economic, social and environmental health aspects of contaminated land remediation [27]. A holistic approach for the management of contaminated land should ideally include an assessment of the environmental risk of the contamination, an assessment of the environmental, social and health impact of the remediation process and a cost-benefit analysis of the remediation project [27].

3.1 Ecosystem services to define the goal of the remediation project

The relative sustainability of a soil remediation project also depends on the objectives of the remediation [28], so that it is not possible to give an overall ‘sustainability score’ to a specific soil remediation technique. More and more, the concept of ‘ecosystem services’ is taken as a starting point to determine the goal of a soil remediation project. Ecosystem services can be defined as the resources and processes that are supplied by natural ecosystems. Cleanup activities can modify ecosystem services, so the determination of ecosystem services is important to avoid unwanted negative effects of soil remediation operations. Ecosystem health is quantified in LCA by linking inventory systems to so-called impact categories. During the last decade, developments in Life cycle analysis have resulted in the definition of so-called mid-point impact categories, such as ecotoxicity and global warming potential. These indicators are characterized by a relatively low uncertainty and a relatively easy interpretation. On the other hand, end point indicators such as biodiversity have to deal with a lot of uncertainty, which should also be accounted for in LCA [29].

3.2 Aspects with regard to human health risk assessment and ecotoxicity

Besides the estimation of contaminant emissions (data inventory faze), fate and exposure modeling and the assessment of ecotoxic effects are essential issues. At the moment, there is no consensus regarding soil quality in the EU. A variety of risk assessment tools is used to assess the risk of contaminated soils towards human health and ecosystem health [30].

Initially, the assessment of the global impact on terrestrial ecosystems within life cycle impact assessment was limited to the impact on soil organisms only because terrestrial vertebrates require a complex modeling that requires data on toxicity upon terrestrial vertebrates and the multiple exposure pathway that should be considered [31]. During the last few years, several impact assessment methods for human health and ecotoxicity have been developed. Pizzol et al. [32] give an overview of different impact assessment methods of heavy metals on human health, but conclude that LCA practitioners should chose each model which has the highest consensus with regard to their specific problem. In the US, the USEtox model is recommended for human and eco-toxicity LCIA, although its uncertainties are still large and it has some limitations as it doesn't consider some metals and is not integrated in any complete LCIA method.

For organic contaminants, changes in concentrations over time, such (bio)degradation, volatilization,… are taken into account in risk assessments. For heavy metals, total concentrations in soil are mostly used as an input in risk assessment models. However, heavy metals in contaminated soil are rarely released completely, as only a portion of heavy metals is "bio-available" or "geo-available" (which means that the metals can be released and become available for biological uptake).

For remediation projects for soils and sediments, this is of course an important consideration, since only a portion of the total metal load in soils or sediments can be considered as 'geo-available' or mobile. In a risk-conservative approach, it is assumed that all the metals contained in a solid matrix (soil, sediment, waste material,….) will be released. When this 'total metal content' is used as an input in LCA analysis, this will most likely result in an overestimation of the risk associated with the heavy metals. Additionally, LCA typically covers an extended period of time (depending on the life cycle of a product or process), so long-term heavy metal emissions have to be assessed. Te assessment of heavy metal release from soils and sediments on the long term is still controversial. Several methods and procedures have been proposed to estimate the long term emissions of heavy metals contained in soils, sediments and waste materials, but there is no consensus on which method performs best [33].

3.3 Financial aspects and financial risks of soil remediation

In most remediation projects, a cost-benefit analysis is already carried out. However, the benefits for the environment and human health are well quantified in LCA, but not always expressed in monetary terms and thus not systematically included in the cost-benefit analysis.

Moreover, certain soil remediation technologies, especially the more 'gentle' remediation technologies are characterized with a lot of uncertainty with regard to the time frame in which the final remediation goals will be achieved. Unexpected situations can result in an increase of the costs of the remediation project. This uncertainty associated with soil remediation projects makes it difficult to make a good estimation of the cost of the remediation. Contaminated sites carry great financial risk, and potentially a great liability for their owners and investors toward authorities. Therefore, there is a need for practical tools that help practitioners in choosing the correct technology that will not only be effective but also will minimize the financial risk associated with the cleanup.

3.4 Involvement of stakeholders

In the past, remediation techniques have mostly been selected without consulting the many stakeholders that are involved in soil remediation techniques such as, for example, local communities affected by the contaminated land and its related remediation operations. "Interactions among all disciplines and stakeholder groups are essential to forge partnerships that will solve environmental problems, rather than deal with only one aspect in isolation" [34].

Besides addressing the environmental impact of the remediation activities for a specific site, attention should also be paid to the consequence of reintroducing a remediate site into the economy [25]. Brownfield development and the displacement of economic activities from suburban to urban sites can counteract the phenomenon of urban sprawl.

From a legal point of view, there should be ways to encourage the use of sustainable remediation technologies, together with a disconnection of treated soils from the definition of waste [35].

Finally, the focus should move from remediation, whether sustainable or not, to prevention of soil contamination [36]. Some countries (for example Flanders (Belgium)) have already adapted their environmental legislation, putting greater emphasis on prevention, but a lot is still expected from innovative technological solutions.

References

[1] EEA 2007. Europe's Environment. The fourth assessment. Chapter 2: Environment and health and the quality of life. EEA, Copenhagen, 2007, 441 pp.

[2] Sorvari, J., Antikainen, R., Kosola, M.J., Hokkanen, P., Haavisto, T., Eco-efficiency in contaminated land management in Finland-Barriers and

development needs. *Journal of Environmental Management,* **90**, pp. 1715-1727, 2009.

[3] Vik, E.A., Bardos, P., Brogan, J., Edwards, D., Gondi, F., Henrysson, T., Jensen, B.K., Jorge, C., Mariotti, C., Nathanail, P., Papssiopi, N., Towards a framework for selecting remediation technologies for contaminated sites. *Land Contamination and Reclamation*, **9(1),** pp. 119-127, 2001.

[4] Vermeulen, H.J., van Veen, H.J., van Reijsen, I.E.R., 2006. Working together in research and development for sustainable land management in Europe.07.06.2006. Snowman vision paper, 23 pp., 2006.

[5] Harbottle, M. J., Al-Tabbaa, A., Evans, C. W., Assessing the true technical / environmental impacts of contaminated land remediation - a case study of containment, disposal and no action. *Land Contamination and Reclamation*. **14 (1)**, pp. 85-99, 2006.

[6] Onwubuya, K., Cundy, A., Puschenreiter, M., Kumpiene, J., Bone, B., Greaves, J., Teasdale, P., Mench, M., Tlustos, P., Mikhalovsky, S., Waite, S., Friesl-Hanl, W., Marschner, B., Ingo Müller. Developing decision support tools for the selection of "gentle" remediation approaches .*Science of The Total Environment*, **407(24),** pp. 6132-6142, 2009.

[7] Tukker, A., Life cycle assessment as a tool in environmental impact assessment. *Environmental Impact Assessment Review*, **20(4),** pp. 435-456, 2000.

[8] Hendrickson, C., Lave, B.L., Matthews, H.S., Environmental Life Cycle *Assessment of Goods and Services: An Input-output Approach*. Resources for the Future Press (RFF Press), 260 pp., 2006

[9] Klemeš, J., Pierucci, S., Worrell, E., Sustainable processes thorough LCA, process integration and optimal design. *Resources Conservation and Recycling*, **50(2)**, pp. 115-121, 2007.

[10] Diamond, M., Page, C., Campbell M., McKenna S. & Lall R. Life-cycle framework for assessment of site remediation options: method and generic survey. *Environmental Toxicology and Chemistry,* **18 (4),** pp. 788-800, 1999.

[11] Suer, P., Nilsson-Påledal, S., Norrman, J., LCA for site remediation: a literature review. *Soil and Sediment Contamination*, **13**, pp. 415-425, 2004

[12] Toffoletto, L., Deschênes, L, Samson, R., LCA for ex-situ bioremediation of diesel-contaminated soil. *International Journal of Life Cycle Assessment*, **10(6),** pp. 406-416, 2005.

[13] Lesage, P., Ekvall, T., Deschênes, L., Samson, R., Environmental assessment of brownfield rehabilitation using two different life cycle inventory models. Part 2: Case Study. *International Journal of Life Cycle Assessment*, **12(6),** 497-513, 2007.

[14] Cadotte, M., Deschênes, L., Samson, R.., Selection of a remediation scenario for a diesel- contaminated site using LCA. *International Journal of Life Cycle Assessment*, **12(4),** pp. 239-251, 2007.

[15] Cappuyns, V., Bouckenooghe, D., Van Breuseghem, L., Van Herreweghe, S., Can thermal soil remediation be sustainable? A case study for a site with a L-NAPL contamination. Accepted for publication in *Journal of Integrative Environmental Sciences*, 2011.

[16] Suer, P., Andersson-Sköld Y., Biofuel or excavation? - Life cycle assessment (LCA) of soil remediation options. *Biomass and Bioenergy*, **35(2),** pp. 969-981, 2011.

[17] Volkwein, S., Hurtig, H.W., Klöppfer, W., Life Cycle Assessment of Contaminated Sites Remediation. *International Journal of Life Cycle Assessment*, **4(5),** pp. 263–274, 1999.

[18] Blanc, A., Métivier-Pignon, H., Gourdon, R., Rousseaux, P., Life cycle assessment as a tool for controlling the development of technical activities: application to the remediation of a site contaminated by sulfur. *Advances in Environmental Research,* **8(1-3),** pp. 613-627, 2004.

[19] Beinat, E., van Drunen, M.A., Janssen, R., Nijboer, M.H., Koolenbrander, J.G.M., Okx, J.P., Schütte, A.R., *The REC decision support system for comparing soil remediation options; A methodology based on Risk reduction, Environmental merit and Costs.* September 1997, CUR/NOBIS, 93 pp., 1997.

[20] Page, C., Diamond, M., Campbell, M., McKenna, S., Lifecycle framework for the assessment of site remediation options: case study. *Environmental Toxicology and Chemistry*, **18**, pp. 801-810, 1999.

[21] Vignes, R.P., Limited Life Cycle Analysis: A tool for the environmental decision-making toolbox. *Strategic Environmental Management*, **1(4),** pp. 297-332, 1999.

[22] Godin, J., Hainsc, J.F. Deschénesa, L., Samson, R., Combined Use of Life Cycle Assessment and Groundwater Transport Modeling to Support Contaminated Site Management. *Human and Ecological Risk Assessment: An International Journal*, **10(6),** pp.1099–1116, 2004.

[23] Hellweg, S., Hofstetter, T.B., Hungerbuëhler, K., Time-dependent life-cycle assessment of slag landfills with the help of scenario analysis: the example of Cd and Cu. *Environmental Impact Assessment Review*, **20(4),** pp. 435-456.

[24] Bayer, P. & Finkel, F., Life cycle assessment of active and passive groundwater remediation technologies. *Journal of Contaminant Hydrology* **83,** pp. 171–199, 2006.

[25] Lesage, P., Ekvall, T., Deschênes L., Samson, R., Environmental assessment of brownfield rehabilitation using two different life cycle inventory models. Part 1: methodological Approach. *International Journal of Life Cycle Assessment*, **12(6),** pp. 391-398, 2007.

[26] Critto, A. & Agostini, P., Using multiple indices to evaluate scenarios for the remediation of contaminated land: the Porto Marghera (Venice, Italy) contaminated site. *Environmental Science and Pollution Research*, **16**, pp. 649–662, 2009.

[27] Pollard, S.J.T., Brookes, A., Earl, N., Lowe, J., Kearney, T., Nathanail, C.P., Integrating decision tools for the sustainable management of land contamination. *Science of the Total Environment* **(1-3),** pp. 15-28, 2004.

[28] Harbottle, M., Al-Tabbaa, A., Evans, C.W., Sustainability of Land Remediation. Part 1: Overall Analysis, Proceedings of the Institution of Civil Engineers – *Geotechnical Engineering*, **161 (2),** pp. 75-92.

[29] Bare, J.C., Hofstetter, P., Pennington, D.W., de Haes, H.A.U., Life Cycle Impact Assessment Workshop Summary. Midpoints versus Endpoints: The Sacrifices and Benefits. *International Journal of Life Cycle Assessment*, **5(6),** pp. 319 – 326, 2000.

[30] Bone, J., Head, M., Barraclough, D., Archer, M., Scheib, C., Flight, D., Voulvoulis, N., Soil quality assessment under emerging regulatory requirements. *Environment International*, **36(6),** pp. 609-622, 2010.

[31] Haye, S., Slaveykova, V.I., Payet, J., Terrestrial ecotoxicity and effect factors of metals in life cycle assessment (LCA). *Chemosphere,* **68**, pp. 1489–1496, 2007.

[32] Pizzol, M., Christensen, P., Schmidt, J., Thomsen, M., Eco-toxicological impact of "metals" on the aquatic and terrestrial ecosystem: A comparison between eight different methodologies for Life Cycle Impact Assessment (LCIA), *Journal of Cleaner Production*, In Press, 2011

[33] Pettersen, J., Hertwig, E.G., Critical Review: Life cycle inventory procedures for long-term release of metals. Environmental Science and Technology, **42(13)**, pp. 4639-4647, 2008.

[34] Burger, J., Environmental management: Integrating ecological evaluation, remediation, restoration, natural resource damage assessment and long-term stewardship on contaminated lands. *Science of the Total Environment,* **400(1-3),** pp. 6-19, 2008.

[35] Wallace, S., Sustainable waste management in contaminated land redevelopment. *Land Contamination and Reclamation*, **12(2),** pp. 101-107.

[36] Gilmore I., A critique of soil contamination and remediation: the dimensions of the problem and the implications for sustainable development. Bull Sci Technol Soc. 21: 394-400, 2008.

Study of bisphenol A in sanitary landfill soil

N. C. Vieceli, E. R. Lovatel, E. M. Cardoso & I. N. Filho
Center of Exact Sciences and Technology, Institute of Environmental Sanitation, University of Caxias do Sul, Brazil

Abstract

The purpose of this work was the quantitative analysis of bisphenol A (BPA), in soil of two sanitary landfills in Caxias do Sul and Farroupilha, RS-Brazil. The samples were collected from closed cells and extracted by Soxhlet. The organic extracts were analyzed by Gas Chromatography with Flame Ionization Detector (GC/FID) and BPA was quantified by the internal normalization method. Standard bisphenol A (Sigma-Aldrich, 99.00%) was used for qualitative and quantitative analysis. The average concentration of BPA was of 21.30 (±0.61) $\mu g\ g^{-1}$ of soil. The BPA concentration found in this work is much higher than the others reported in the literature. This result is worrying because there is a suspicion that bisphenol A disrupts the endocrine system.
Keywords: bisphenol A, sanitary landfill, gas chromatography.

1 Introduction

Around 15% (in volume) of the municipal solid wastes buried in both municipal landfills are polymeric materials. Depending on certain soil conditions, like humidity and pH, these polymeric materials can release a relatively high quantity of organic compounds with toxicological potential called plasticizers.

Plasticizers are organic esters added to polymers to facilitate processing and to increase flexibility and toughness of the final product by internal modification of the polymer molecule. Due to their use in packaging, clothes, films, paints, adhesives, cosmetics, ink printers and many other products, plasticizers compounds are widespread in all environments. Several plasticizers were detected in mineral, ultra pure and tap waters [1], municipal solid waste compost

www.witpress.com, ISSN 1743-3541 (on-line)
doi:10.2495/CHEM110211

[2], sewage and wastewater treatment sludge [3], river sediments [4], landfill leachate [5–7] and swine slaughterhouse wastewaters [8], among other environmental samples.

One of the main forms of exposure to plasticizers is through food, in addition to medical materials and by occupational contact [9]. Among the problems that can be caused by plasticizer compounds are: reproductive anomalies (morphological and functional gonadal dysfunction, e.g., infertility and decreased libido) and congenital malformations [10, 11]. The plasticizers can also influence the mobility and bioavailability of toxic substances like polychlorinated biphenyls and metal ions by changes in water or lipid solubility [4].

Bisphenol A (BPA), 4,4′-(1-methylethylidene)bisphenol (CAS nº 80-05-7) is largely applied as a monomer in the production of polycarbonate and some epoxy resins [12]. BPA improve some physical properties of the polymeric material, like resistance, hardness, and thermal stability. The ether bonds in the polymer also increase the chemical resistance. Figure 1 shows the chemical structure of BPA.

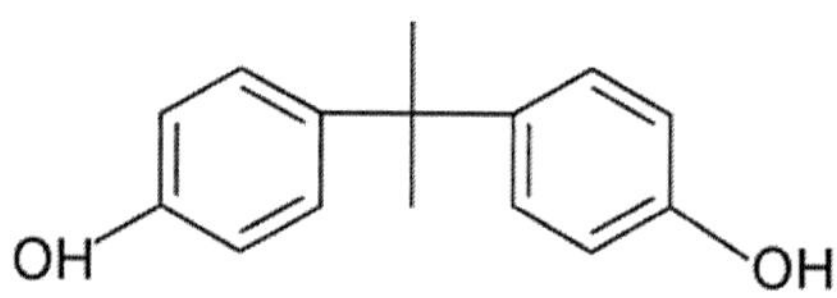

Figure 1: Chemical structure of BPA.

Epoxy resins are commonly employed as adhesives and cover material. The environmental persistence of BPA can be due to the low vapor pressure, moderate water solubility and partition coefficient organic carbon/water. In the environment, most of the BPA (78 to 99.3%) is fixed in the soil. BPA can interfere in the functioning of the endocrine system, in the hormones' action mechanism. They are called endocrine deregulators or endocrine disruptors [5, 13]. BPA can also interfere in the action of steroid hormones inhibiting the proteins responsible for the liver metabolism [14].

In sanitary landfills, the BPA leached from the wastes is mixed with the leachate formed by the microbiological degradation of the organic content of refuse. Due to the action of rainfall, the leachate goes down through the landfill, and can contaminate the soil, surface and groundwater [15]. Typically the plasticizer concentrations in the soil are between 30 and 50 $\mu g\ g^{-1}$ [4].

Despite the wide utilization and environmental relevance of BPA there is a lack of publications dealing with the study of this compound in soil. Therefore, the main objective of this work is the quantitative analysis of bisphenol A in samples of two landfill leachate soils, using Soxhlet extraction and Gas Chromatography with Flame Ionization Detector.

2 Materials and methods

2.1 Samples

Five soil samples (500 g each) were collected from the C10 cell of the São Giácomo sanitary landfill (Caxias do Sul, RS, Brazil) and one from the Farroupilha sanitary landfill (Farroupilha, RS, Brazil). Both cells were closed in 2003. The samples were collected from a depth of 2 m below the landfill cover in glass flasks (500 mL) with the cap covered by aluminum foil. The flasks were previously washed with n-hexane and distilled water and dried at 300°C. The same purification procedures were adopted with all the glass material that had direct contact with the samples. The soil particle size was homogenized in a steel sieve with 2 mm mesh. The soil was stored at 4°C, protected from light until extraction (no more than 24 h after collection).

2.2 Reagents and solvents

All the reagents and solvents were of p.a. grade and, when necessary, distilled twice in a glass apparatus.

2.3 Physical-chemical analysis

The main parameters tested were: pH (potentiometric method in humid base), moiety (gravimetry), total nitrogen (Kjeldahl and titrimetric methods), total carbon (Walkley-Black method) and phosphorous (humid digestion and colorimetry).

2.4 Sample extraction

After soil particle size homogenization, six soil aliquots with an average mass of 10.32 g (±0.24) were extracted by Soxhlet in 125 mL of *n*-hexane by 4 h. After this, the organic extracts were dried until constant mass and re-diluted with *n*-hexane (1.0 mL) and analyzed by GC/FID. Throughout the extraction process, no plastic or rubber materials were allowed to come into contact with the samples.

2.5 Instrumental analysis

The organic extracts were analyzed by GC/FID in a PerkinElmer chromatograph (model Autosystem) equipped with a capillary column elite plot (30 m x 0.25 mm x 0.25 μm). The analysis conditions are shown in Table 1.

2.6 Recovery grade and detection limit

The recovery grade for BPA was measured by the extraction of a standard solution (20 μg mL^{-1}). The detection limit was evaluated by the analysis of successive dilutions of a standard solution at 100 μg mL^{-1}.

Table 1: Instrumental analysis conditions.

Parameter	Condition
Initial temperature	100°C (hold 10 min)
Heating ramp	10°C/min
Final Temperature	250°C (hold 5 min)
Carrier gas	He
Flux	1 mL/min
Injection mode	Splitless
Injection volume	1 µL

3 Results and discussions

The BPA was not detected in the Farroupilha sanitary landfill sample, suggesting that different residues were buried in the landfills. Thus, our discussion will be focused on the São Giacomo sanitary landfill samples.

Figure 2 shows the chromatogram of the extraction blank where no BPA peak was observed and a typical chromatogram of the real samples organic extract.

The recovery grade of BPA was 98.73% (±4.51). The detection limit was 0.5 µg mL^{-1} and the detector response was linear between 0.5 and 25 µg g mL^{-1}

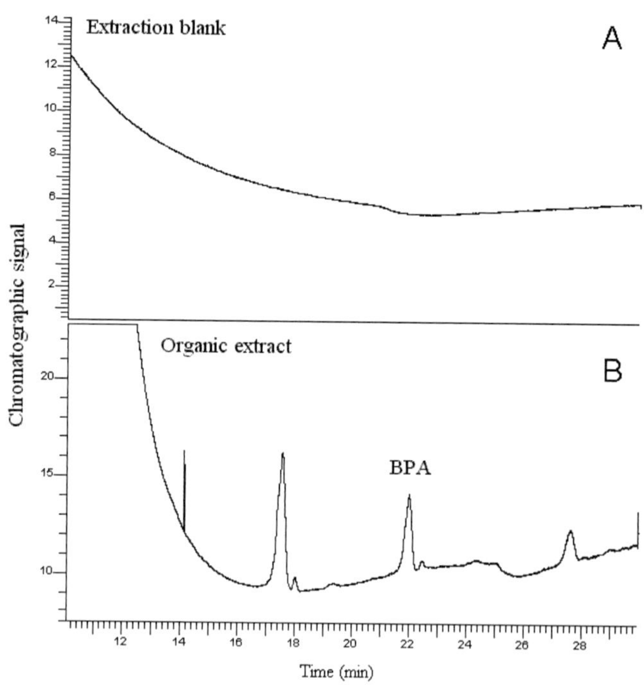

Figure 2: Chromatogram of the extraction blank (A) and the organic extract (B).

(R^2 = 0.9184; equation: y = 3.4968x – 0.3898). The average BPA concentration in the samples was 21.30 (±0.61) $\mu g\ g^{-1}$. Yamamoto et al. [12] had reported average BPA concentrations of 0.27 $\mu g\ mL^{-1}$ in hazardous waste landfill samples. Thus, a landfill that receives only municipal solid wastes, like the one studied in our work can be considered much more hazardous with regard to the underground water contamination by BPA, than a typical hazardous waste landfill.

Samples of soil located outside the São Giacomo landfill (where the residues were not buried) show an average BPA concentration of 3.16 (±0.14) $\mu g\ g^{-1}$, i.e. far below that found in the closed cells. This reinforces the hypothesis that the high concentration of BPA can be explained by the high content of polymeric material buried in the São Giácomo landfill (Figure 3). The non observance of residue segregation (into organic and inorganic categories) could be the source of the high BPA concentration.

Figure 3: Polymeric material buried in the São Giácomo landfill.

The physical-chemical and microbiological characteristics of the landfill soil can enhance the release of BPA from plastic residues The results of the physical-chemical parameters (Table 2) show that both cells are in an unstable methanogenic phase [16]. In this phase the fermentation and hydrolysis of the residues generates, mainly, intermediary organic volatile acids. The leachate generated by this fermentation reduces the soil pH. Thus, at the pH found for our samples (Table 2), the BPA can be poorly ionized and must be adsorbed in the organic matter. This hypothesis is reinforced by the high organic matter content of the soil samples (Table 2).

Table 2: Physical-Chemical characterization of the soil samples.

Parameter	Result	Method
pH	5.24	Potentiometric
Humidity (%)	38.7	Gravimetric
Nitrogen (%)	0;3	Kjeldahl – Titrimetric
Phosphorus (%)	<2.8	Humid digestion - Colorimetric
Organic carbon (%)	16	Walkley-Black

Therefore, the low pH and high organic content conditions of the soil conjugated with the presence of plastic material in the municipal solid wastes, created a favorable environment for the BPA release into the sanitary landfill.

4 Conclusions

The BPA concentration in the in the soil of the São Giacomo landfill is far above that found in leachate of hazardous wastes. This is a worrying result since this contaminant can migrate to the superficial and underground waters, near the sanitary landfills. BPA is a pollutant suspected of estrogen activity. It is very reasonable to assume that the sources of most of the BPA found in this sanitary landfill are non segregated domestic wastes. Therefore, it is extremely important to re-link environmental education programs focusing on the segregation of domestic wastes in inorganic, organic and recyclable categories. Likewise, the environmental laws have to be reviewed and expanded to include the plasticizers concentrations in the environment as a parameter for the soil quality and the treatment of sources of potential water pollution.

Acknowledgement

CNPq, FAPERGS, CODECA – Caxias do Sul Development Company, Jobel Engineering, Procel and Farroupilha Ambiental Inc.

References

[1] Akon, H., Yoon, B., Takayuki, K., Mariko, H., Maya, M. & Takashi, N., Separation of endocrine disruptors from aqueous solutions by pervaporation: Dioctylphthalate and butylated hydroxytoluene in mineral water. *Journal of Applied Polymer Science*, **94(4)**, pp. 1737–1742, 2004.

[2] Gonzalez-Vila, F.J., Saiz-Jimenez, C. & Martin, F., Identification of free organic-chemicals found in composted municipal refuse. *Journal of Environmental Quality*, **11**, pp. 251–254, 1982.

[3] Boyd, S.A., Sommers, L.E. & Nelson, D.W., Infrared spectra of sewage sludge fractions – evidence for an amide metal-binding site. *Soil Science Society of America.* **43**, pp. 893–899, 1979.

[4] Möder, M., Popp, P. & Pawliszyn, J., Characterization of water-soluble components of slurries using solid-phase microextraction coupled to liquid chromatography mass spectrometry. *Journal of Microcolumn Separation*, **10(2)**, pp. 225–234, 1998.

[5] Bauer, M.J. & Herrmann, R., Estimation of the environmental contamination by phthalic acid esters leaching from household wastes. *Science of the Total Environment*, 208 **(1–2)**, pp. 49–57, 1997.

[6] Nascimento, I.F, von Muhlen, C., Schossler, P. & Caramão, E.B., Estudo de compostos orgânicos em lixiviado de aterros sanitários por EFS E CG/EM. *Química Nova*, **24(4)**, pp. 554–556, 2001. Abstract in English.

[7] Nascimento, I.F., von Muhlen, C., Schossler, P. & Caramão, E.B., Identification of some plasticizers compounds in landfill leachate. *Chemosphere*, **50**, pp. 657–663, 2003.

[8] Aguilar, C.P., Peruzzolo, M., Di Luccio, M., Dallago, R.M. & Nascimento, I.F., Qualitative study of organic compounds in wastewaters of a swine slaughterhouse. *Environmental Monitoring and Assessment*, **116**, pp. 103–110, 2006.

[9] Wams, T.J., Diethylhexylphthalate as an environmental contaminant: a review. *Science of the Total Environment*, **66**, pp. 1–16, 1987.

[10] Nelson, P., Epidemiology, biology, and endocrine disrupters. *Occupational and Environmental Medicine*, **60**, pp. 541–542, 2003.

[11] Mantovani, A., Problems in testing and risk assessment of endocrine disrupting chemicals with regard to developmental toxicology. *Chemosphere*, **39**, pp. 1293–1300, 1999.

[12] Yamamoto, T.; Yasuhara, A.; Shiraishi, H. & Nakasugi, O., Bisphenol A in hazardous landfill leachates. *Chemospher*e, **42**, pp. 415–418, 2001.

[13] Baird, C., *Environmental Chemistry*, W.H. Freeman and Company: New York, 1999.

[14] Goloubkova, T.S. & Spritzer, P. M., Xenoestrogênios: o Exemplo do Bisfenol-A. *Arquivos brasileiros de Endocrinologia e Metabologia*, **44**, pp. 323–330, 2000. Abstract in English.

[15] Keenan, J.D., Steiner, R.L. & Fungaroli, A.A., Landfill leachate treatment. Journal of Water Pollution Confederation, **56 (1)**, pp. 27–33, 1984.

[16] Rees, J.F., The fate of carbon compounds in the landfill disposal of organic matter. Journal of Chemical technology and Biotechnology, **30**, pp. 161–175.1980.

Thread as a substrate for low-cost point-of-care diagnostics

X. Li, D. Ballerini, J. Tian &W. Shen
Australian Pulp and Paper Institute,
Department of Chemical Engineering, Monash University, Australia

Abstract

This study presents new applications of thread-based microfluidic systems in chemical and biochemical diagnostics. Our recent reports have demonstrated the capability of fabricating low-cost microfluidic devices using ubiquitous multifilament materials such as threads. The gaps between fibres in threads provide capillary wicking channels for liquid transport; therefore, liquid can penetrate along threads without the need of external forces. The new thread-based system, combined with the cheap dosing tools, is capable of easily and rapidly semi-quantifying the concentration of analytes in human body fluids. The utility of this system is further extended for the rapid and easy blood grouping with only small amount of whole blood (~ 2 μL for ABO blood typing), which is an essential test before blood transfusion. These low-cost and portable microfluidic devices are easy to fabricate, simple to use, and do not require powers such as electricity; thus providing a desirable analytical platform for point-of-care applications.
Keywords: low-cost diagnostics, microfluidic, point-of-care, thread-based.

1 Introduction

Modern microfluidic systems have been designed to carry out complicated analytical measurements for a range of applications [1]. However, such sophisticated microfluidic systems are usually expensive and the users always need to be trained to acquire the operation skills, thus restricting the practicability of these systems. The increasing needs of low-cost microfluidic devices which are readily accessible to average users, especially those people from under-developed areas or remote regions compel the development of new

WIT Transactions on Ecology and the Environment, Vol 154, © 2011 WIT Press
www.witpress.com, ISSN 1743-3541 (on-line)
doi:10.2495/CHEM110221

concepts and designs of inexpensive and easy-to-use microfluidic devices [2]. Paper-based microfluidic device, as a novel concept, was firstly proposed in 2007, then a plenty of subsequent studies from several research groups further developed this concept, either in the fabrication methods [3–7] or in the new applications [7-9]. Most recently, our group reported that cotton thread is also an ideal material to create microfluidic systems [10].

As an ancient and widely used material, cotton thread is cheap and can be manufactured in almost all areas of the world, even in the under-developed areas. Natural cotton fibres are hydrophobic because of the wax on the fibre surface [11]. However, with appropriate treatment (e.g. plasma), cotton thread can be made hydrophilic, and the gaps between thread fibres provide the capillary channels for liquids transport along thread without external pumps or driving forces. Besides, cotton thread is white and, therefore, presents a suitable background for colour expression for the colorimetric assay. All these properties make thread a suitable material to fabricate low-cost microfluidic thread-based analytical devices (μTADs) for biomedical or environmental applications.

In this paper we showed the surface modification of cotton thread using plasma for building μTADs and developed a new method to measure the surface energy of thread. Ink jet ink solution was used to analyze liquid penetration and colour expression properties of hydrophilic threads. We reported for the first time that μTADs can measure analyte concentration semi-quantitatively just with the cheap and simple sample dosing tools (e.g. sewing needle). This innovation avoided the use of expensive and precise quantitative sampling tools (e.g. micropipette), which made μTADs more suitable for field use under very limited conditions. Nitrite (NO_2^-), as an important biomarker indicating human health [12], was used as an example analyte to evaluate this analytical capability of μTADs.

Another application of threads in point-of-care diagnostics presented is the identification of blood groups by virtue of separating agglutinated red blood cells (RBCs) from serum based on the sensitivity of the flow resistance of large particles in narrow capillary channels within threads. This new application has high diagnostic value for that the correct grouping of human blood is imperative before blood transfusion and it had been reported that without ABO compatibility testing, around one-third of unscreened blood transfusions would be expected to cause a haemolytic reaction [13].

All those new applications and advantages of μTADs showed that this new system is simple and robust enough for the end users to perform basic healthcare sensing without special training. MicroTADs are portable, disposable, can be easily preserved and maintain a low cost of construction, making them attractive for diagnostics and point of care tests especially in remote and developing regions.

2 Materials and methods

2.1 Source of thread

The raw (unbleached) cotton thread and polyester thread were obtained from the School of Fashion and Textiles, RMIT University, Melbourne, Australia. Our prior work showed that the raw cotton thread is hydrophobic to water because of the wax on the fibre surface. Plasma treatment can make thread hydrophilic, enabling liquid penetration along it.

2.2 Thread surface analysis before and after plasma treatment

An XPS analysis with an AXIS HIS spectrometer (Kratos) and monochromatized Al Kα radiation was used to evaluate the surface modification of cotton thread with plasma. Thread samples were wound around a sample holder to form a closely packed mesh providing the necessary area (~2 mm in diameter) for the X-ray beam. Whatman no. 1 filter paper was used as the cellulose control. The photoelectron emission angle was 90° with respect to the sample surface, corresponding to a maximum sampling depth of ca. 10 nm.

2.3 Preparation of test solutions

Water (Millipore, 18 MΩ) was used for making all solutions in this study. The commercial Canon ink jet magenta ink (http://www.canon.com.au/) was defined as the 100% ink solution; and it was diluted with water to obtain the 25%, 6.25% and 1.56% ink solutions for the colour display experiment on the hydrophilic thread.

The colorimetric detection of NO_2^- is based on the principle of the Griess reaction [14]. The stock solution of NO_2^- (10.0 mmol/L) was prepared by dissolving 69.0 mg sodium nitrite (≥99%) in 100 mL water. Then this stock solution was diluted with water to get serially diluted NO_2^- standard solutions with the concentrations of 1000, 500, 250, 125 and 0 μmol/L. The indicator solution for NO_2^- contains 50 mmol/L sulfanilamide (≥99%), 330 mmol/L citric acid (≥99.5%), and 10 mmol/L N-(1-naphthyl) ethylenediamine (≥98%) [12]. All of the chemicals were purchased from Sigma-Aldrich.

Epiclone™ anti-A, anti-B and anti-D monoclonal grouping reagents were obtained from CSL Australia. Blood samples (A+ and O-) were sourced from donors of known blood type, with samples being drawn by a trained nurse, and stored in Vacutainer® test tubes containing lithium-heparin anticoagulant. All blood samples were stored at 4°C, and used within 5 days of withdrawal.

2.4 Preparation of μTADs for semi-quantitative detection of NO_2^-

Threads were used with polymer film to fabricate μTADs by simply sewing the hydrophilic cotton thread through the film; colorimetric assays can be performed on threads. The white solid line (Figure 1) represents the stitches on the front side of the polymer film, while the dashed lines are the stitches on the back side.

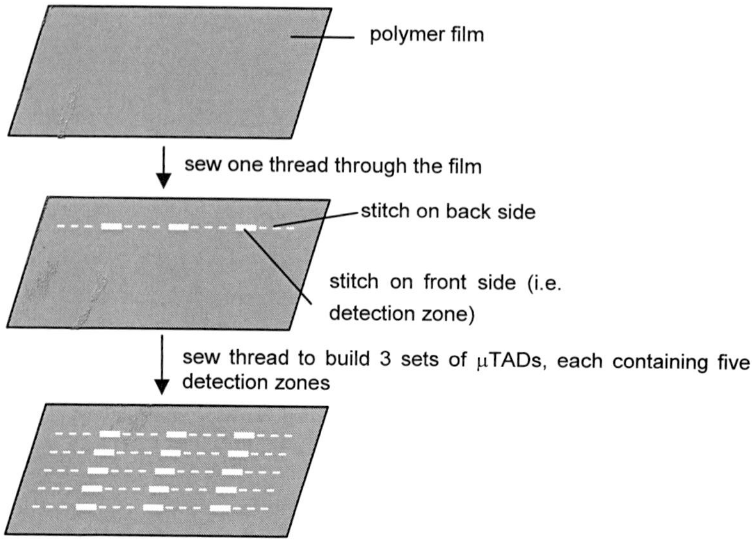

Figure 1: The scheme of the fabrication of μTADs.

Each front side stitch (~3 mm) was used as a detection zone of μTADs. To quantify the assay results, μTADs were scanned using a desktop scanner (Epson Perfection 2450 PHOTO, colour photo setting, 1200 dpi) and the images were analyzed using ImageJ 1.41o. For any selected area (i.e. detection zone), ImageJ calculated the average gray value within the selection, and background-corrected data was obtained by subtracting this value from the average intensity of the detection zone containing blank control solution (0 μmol/L NO_2^-), thereby showing the semi-quantitative results which are correlated with the concentration of the sample solutions.

2.5 Preparation of thread-based platform for blood grouping

Polyester threads were treated with antibody via soaking in the grouping reagent followed by blotting with standard blotting paper (Drink Coster Blotting, 280 $g \cdot m^{-2}$) to remove any surplus of antibody solution from the thread. After drying under fume hood for 10 min, coated threads were ready to be used for ABO and Rh blood grouping.

3 Results and discussion

3.1 Hydrophobic-hydrophilic contrast of untreated and plasma-treated threads

Nisbet et al. have shown that the polar liquid penetration into a porous material occurred when the surface tension (γ) of the liquid was lowered to a critical value and this value could be considered as the approximate surface free energy of the porous material [15]. Based on this principle, a series of isopropanol (IPA)-water solutions were prepared with different IPA mass percentages of 16, 18, 20, 22, and 24% (surface tension values were summarized in Table 1 [16]), then the surface free energy of natural cotton thread was measured by depositing a drop of IPA-water solution of different concentrations on thread and observing the drop penetration behaviour (i.e. the liquid forms a bead on thread surface or penetrates into thread). Table 1 shows that liquids of surface tension higher than 32.4 mN/m cannot penetrate along the thread, while liquids of surface tension lower than 31.2 mN/m can penetrate into thread. That means the surface free energy of the untreated cotton thread was between 31.2 and 32.4 mJ/m^2.

Table 1: Penetration behaviour of IPA + water on cotton thread.

Isopropanol mass %	Surface tension (γ/mN·m^{-1}) at 20° C	Penetration on cotton thread (Yes/No)
16	34.1	No
18	32.4	No
20	31.2	Yes
22	29.9	Yes
24	29.2	Yes

Though natural cotton thread is hydrophobic to water, the fibre surface can be modified into hydrophilic with certain treatment (e.g. plasma in this work). The plasma-treated thread was strongly hydrophilic and easily wettable by water. Such difference in thread's wettability before and after treatment is referred as the hydrophobic-hydrophilic contrast. Figure 2 shows the XPS analysis of the surface chemistry of cotton threads before and after treatment to reveal changes in the surface atomic concentrations of C and O.

Plasma treatment may have two effects on thread surface modification. First, the wax was partially removed, thus exposing the underlying cellulose, which was indicated by the increase in surface concentration of C–O (BE = 286.8 eV). Second, thread surface was substantially oxidized, which was revealed by the increase in surface concentrations of O–C=O, O–C–O and C=O (BE = 289.2, 288.3 and 288.3 eV, respectively). These changes increased the surface oxygen concentration and the surface polarity of the thread, making the plasma-treated thread hydrophilic.

3.2 Liquids penetration along hydrophilic thread and colour display on thread

Liquid penetration along the hydrophilic thread was investigated to understand the relationship between the volume of liquid introduced onto thread and the liquid penetration distance along thread. Four pieces of threads were fixed in parallel onto a supporting film. Magenta ink solutions of different volume (0.1, 0.2, 0.4, 0.8 μL) were introduced onto each thread using a micropipette (Eppendorf Research® 0.1 – 2.5 μL) and penetrated along thread with different distance (Figure 3a). A calibration curve (Figure 3b) from six independent measurements shows that the liquid penetration length almost linearly correlates to the liquid volume (when volume ≤ 0.4 μL). This correlation became curved as the liquid volume increased beyond 0.4 μL. The near-linear section of the calibration curve suggests that the ink "concentration" on the thread (i.e. the amount of dye molecules on each unit length of thread) is constant and is independent on the ink volume. This indicates that a semi-quantitative analysis can still be performed on μTADs even without precise liquid-handling instrument (as long as sample volume ≤ 0.4 μL).

To further prove this prediction, three parallel sets of μTADs were fabricated and each μTAD has four detection zones vertically arranged from top to bottom (Figure 4a). Ink solutions of 0.1 μL of different concentrations (1.56%, 6.25%, 25%, and 100%) were added to μTAD 1 (the left set). Similarly, ink solutions of 0.2 μL were added to μTAD 2 (the middle set) and those of 0.4 μL were added to μTAD 3 (the right set). The adding sequence is shown in Figure 4a. The penetration length of 0.1 μL ink solution is about 5 mm; for ink solutions of

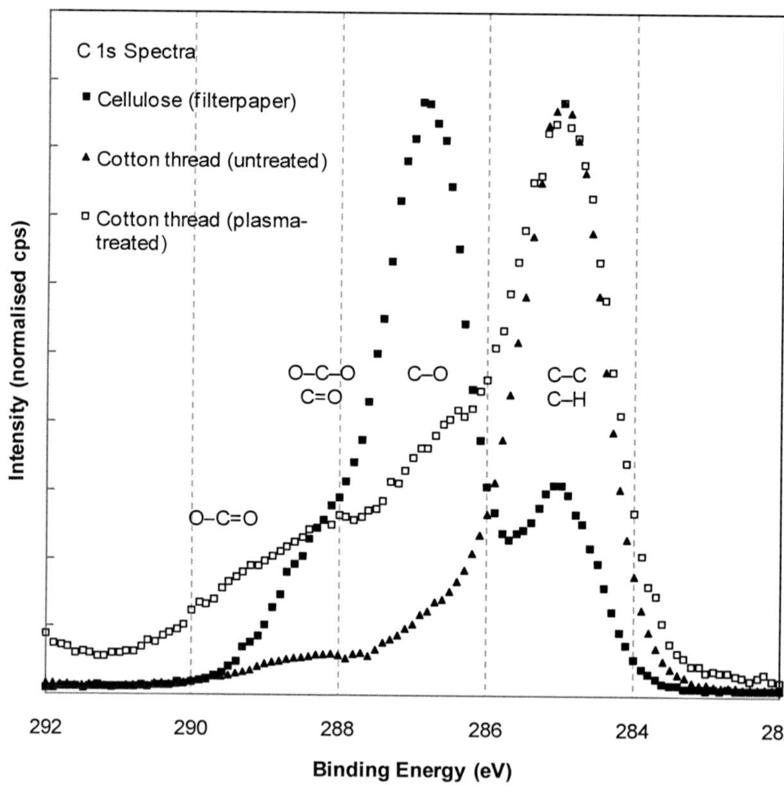

Figure 2: XPS results of the untreated (hydrophobic) thread and plasma-treated (hydrophilic) thread. Filter paper is used as the control sample for cellulose.

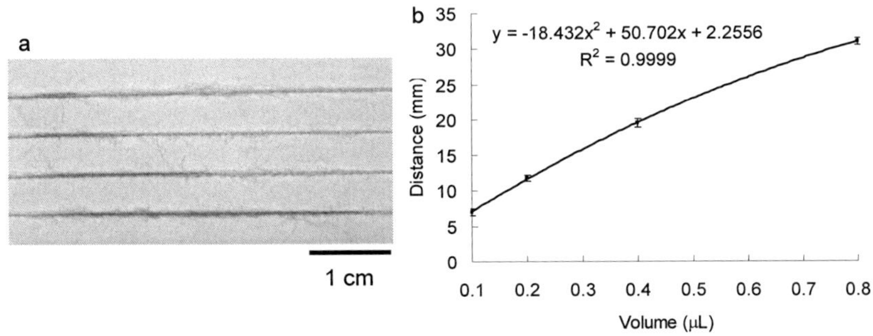

Figure 3: (a) Ink penetration on hydrophilic thread. Ink solutions of different amount (0.1, 0.2, 0.4, 0.8 μL) are introduced onto four pieces of thread from top to bottom; (b) the average ink penetration length obtained from six repeated tests using a Vernier caliper.

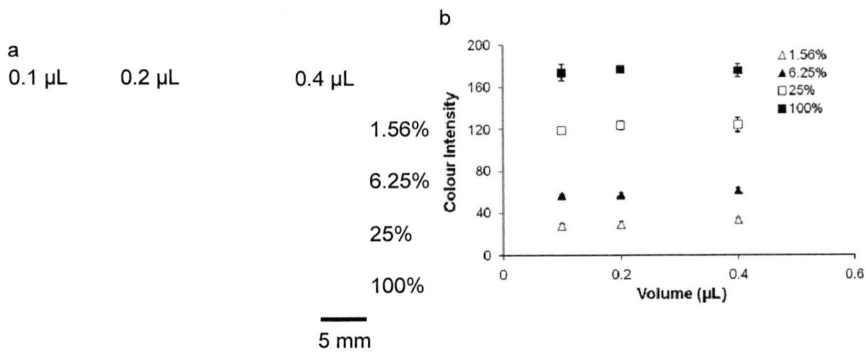

Figure 4: (a) Colour display of different ink solutions on 3 sets of μTADs. The volume of ink is 0.1, 0.2 and 0.4 μL, respectively (from left to right); the concentration of ink is 1.56%, 6.25%, 25% and 100%, respectively (from top to bottom); (b) The average ± s.d. colour intensity of each observation zone is obtained from six independent tests.

larger volume, the extra amount of solution penetrates along thread to the back side of polymer film and cannot be seen. For each μTAD, the colour of ink solution varies from weak to strong as ink concentration increases. However, the colour of ink solution of same concentration and different volume has negligible variation which is proved by colour intensity measurement. For each detection zone, a fixed rectangle area is set to encompass the zone and the mean colour intensity of the selected area is analyzed. Six independent measurements performed provide the average ± standard deviation (s.d.) data of colour intensity (Figure 4b). The results show that the average colour intensity is consistent for ink solution of different volume and same concentration; the small standard deviation indicates that thread is a good substrate for colour display.

3.3 Semi-quantitative measurement of NO_2^- using μTADs and cheap liquid dosing tool

Since the colour intensity of sample solutions spotted on μTADs is not affected by the sample volume (volume ≤ 0.4 μL), semi-quantitative assays can be performed even if high precision liquid-handling instrument is unavailable. A household sewing needle was used in this work as a simple means to spot liquids. We found that by rapidly dipping the needle eye into liquids, it can pick up liquid of 0.22 ± 0.04 μL (average ± s.d., 12 repeats) after moving out from the solution. This small volume variation does not cause a significant colorimetric error for the reasons discussed above. Water and 70% ethanol were used to wash and disinfect the needle eye for repeated use.

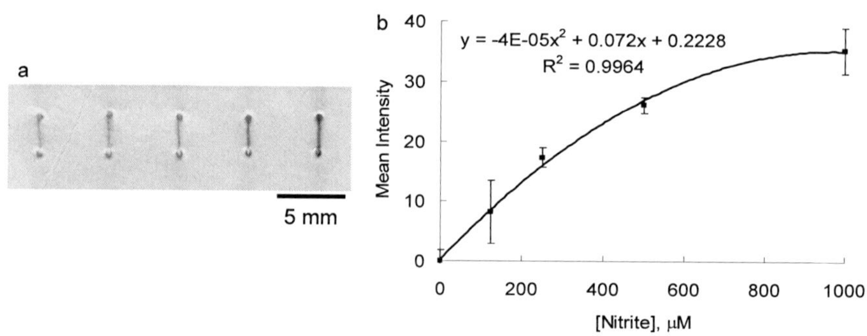

Figure 5: (a) NO_2^- sample solutions of different concentration (0, 125, 250, 500, and 1000 μM) were added using needle eye onto μTAD from left to right. NO_2^- triggered different colour changes on each detection zone; (b) semi-quantitative results of the relationship between NO_2^- concentration and colour intensity (n=6).

Nitrite (NO_2^-) was used as an analyte to demonstrate the feasibility of using μTADs and cheap sample dosing tool to perform semi-quantitative analysis. The hydrophilic thread was soaked in NO_2^- indicator solution and then dried in oven (60°C) for 5 min; the indicator-treated thread was sewn onto a polymer film to fabricate μTAD (procedure in Figure 1) containing five detection zones for colorimetric measurement. Five standard NO_2^- solutions (0, 125, 250, 500 and 1000 μM) were introduced onto each detection zone (from left to right) using the needle eye and caused different colour development from colourless to pink of different shades (Figure 5a). The NO_2^-concentration calibration curve can be obtained through colorimetric analysis of those detection zones with the average ± s.d. data of colour intensity from six independent tests on six μTADs. Therefore, μTADs, together with cheap liquid-handling tool, can be used as new low-cost, low-volume and easy-to-use analytical platform for health or environmental care in developing and remote regions with limited resources; and the users only need to get basic training for utilizing them.

3.4 Blood grouping using thread-based platform

To investigate the feasibility of using antibody-treated threads as blood grouping platform, three types of thread (thread treated with anti-A, anti-B and anti-D, respectively) were immobilized in folded polypropylene films with lodging slits to aid testing, and a micropipette (Eppendorf research® 0.1–2.5 μL) was used to dose whole-blood sample of 1 μL onto each thread. If red blood cells (RBCs) have haemagglutination reaction with the antibody deposited on thread, a separation band can be seen on thread. The separation distance is around 3 mm (similar length to a standard "stitch" found on a shirt cuff), and is therefore easily visible by unaided human eye. Typically, results became observable within 1 min of sample dosing, making the test quite rapid compared to some existing techniques. Identifying a person's ABO and Rh blood type requires three tests, two for A and B antigens to determine ABO grouping, and a further test for D antigen to verify Rh grouping. A results-matrix (Table 2) aids in the interpretation of results.

Table 2: Results-matrix for blood grouping interpretation. Tick indicates a positive result (separation occurs); crosses indicate a negative result (no separation).

<table>
<tr><td>Anti - A</td><td colspan="4">✓ ✓ ✓ ✓</td><td colspan="4">× × × ×</td></tr>
<tr><td>Anti - B</td><td colspan="2">✓ ✓</td><td colspan="2">× ×</td><td colspan="2">✓ ✓</td><td colspan="2">× ×</td></tr>
<tr><td>Anti -D</td><td>✓</td><td>×</td><td>✓</td><td>×</td><td>✓</td><td>×</td><td>✓</td><td>×</td></tr>
<tr><td></td><td>AB+</td><td>AB-</td><td>A+</td><td>A-</td><td>B+</td><td>B-</td><td>O+</td><td>O-</td></tr>
</table>

Identifying the agglutination of RBCs using thread has a unique advantage: large and discrete particles in a suspension system undergo a phase separation from the continuous liquid phase in the narrow capillary channels. As whole-blood sample penetrates through the inter-fibre gaps of the antibody-treated threads, blood serum will dissolve the antibody molecules on the fibre surface. If antibodies bond to the specific binding sites on the antigens of adjacent RBCs (i.e. the haemagglutination reaction happens), the aggregation of RBCs by the antibody molecules leads to the formation of significantly larger particles that cannot be stably suspended in the serum phase. As the sample penetrates along thread, the size of the particles increases and the agglutinated RBCs cannot contribute to the capillary driving force required for the blood sample to continue penetrating forward. Instead, the agglutinated RBCs can only be carried by the serum to move forward. In this situation agglutinated RBCs will be gradually left behind from the serum penetration front. Furthermore, agglutinated RBCs in the inter-fibre channels may act as a "filter" which permits serum to pass, but prevents agglutinated RBCs from passing. This causes the separation of agglutinated RBCs from the serum phase and the separation is visually identifiable by the appearance of a pale pink coloured band between the agglutinated RBCs and dry antibody residue, as seen in Figure 6a, 6c. In the case where the deposited antibody does not react with the antigen of RBCs, no separation is visually detectable (as seen for type O- blood in Figure 6d-f).

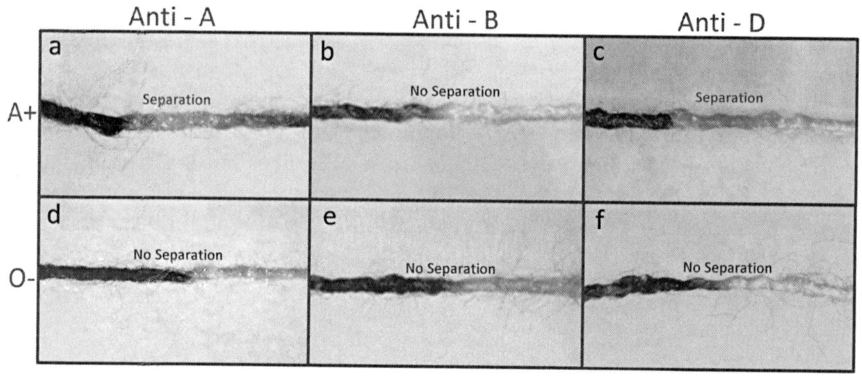

Figure 6: Proof of concept tests with samples of whole blood of type **A+** and **O-** on antibody-treated polyester threads. Columns show results on threads treated with anti-A, anti-B and anti-D from left to right; rows show the results for the different blood types as labelled on the left.

3.5 Storage stability of the devices

Referring to the ultimate practical application of the μTADs, it is important that these devices must be stable for a certain period (e.g. one month or longer) to become commercialized products for consumers. In this study, μTADs fabricated for semi-quantitative NO_2^- detection were stored in a light-tight container in a desiccator at room temperature. The performance of μTADs showed no deterioration after 6 weeks, indicating that these low-cost and easy-to-use μTADs have sufficient shelf life for practical use.

The efficacy of the thread-based platform for blood grouping was investigated after a 4 week storage period. Two sets of antibody-treated threads were stored in microtubes wrapped in foil; one at 4°C the other at ambient temperature. All tests showed the correct grouping, suggesting that the device could be stored for a moderate period of time without degrading if sealed in airtight and light-proof packaging.

4 Conclusion

Thread, a cheap and widely available material, was used in this study to fabricate low-cost microfluidic analytical devices; the analytical capabilities of these devices were studied in detail. The hydrophobic natural cotton threads can be surface modified using plasma treatment into hydrophilic threads, allowing the transport of aqueous liquids via capillary driving force. Threads as liquid transport channels do not rely on physical or chemical barriers, enabling μTADs to transport low surface tension liquids such as organic solvents or oil, which may not be possible for some paper-based microfluidic channels. Threads are white and stainable, making them suitable for color display. Semi-quantitative tests of analytes can be rapidly and easily performed on μTADs through colorimetric reaction using the non-precision household liquid handling items such as sewing needle which are also available in resource-limited regions. Based on this principle the concentration of NO_2^- (a biomarker for indicating some human diseases) in aqueous samples has been successfully measured.

Thread was further used as a porous substrate for blood grouping in this work, which requires only whole blood of small amount (e.g. ~ 2 μL for ABO blood grouping) and eliminates the need for the end user to either handle other testing reagents or perform sample dilutions. Although only ABO and Rh/D blood groups were tested in this study, it is expected that the platform could easily be extended to identify other groups of interest that follow similar antibody/antigen interactions.

All these μTADs and thread-based blood grouping system prototyped are simple and robust enough to be employed by the end user without assistance. They are portable, disposable, can be easily preserved and maintain a low cost of construction, making them attractive for diagnostics and point of care testing especially in remote and developing regions.

Acknowledgements

The authors would like to thank the kind blood donors: Mr. Scot Sharman, Mr. Henri Kröling; as well as Ms. Lisa Collison of the Monash University Health Service for collecting blood donations. The authors also thank Dr. Lijing Wang of the School of Fashion and Textiles, RMIT University, for kindly providing thread samples. The research scholarships of Monash University and the Department of Chemical Engineering are gratefully acknowledged. The Postgraduate Research Travel Grant from Monash Research Graduate School (MRGS) and the New Speakers Contest prize from Appita are also appreciatively acknowledged for supporting the prospective conference participation.

References

[1] West, J., Becker, M., Tombrink, S. & Manz, A., Micro total analysis systems: latest achievements. *Analytical Chemistry*, **80(12)**, pp. 4403–4419, 2008.

[2] Zhao, W. & van den Berg, A., Lab on paper. *Lab on a Chip*, **8(12)**, pp. 1988–1991, 2008.

[3] Martinez, A.W., Phillips, S.T., Butte, M.J. & Whitesides, G.M., Patterned paper as a platform for inexpensive, low-volume, portable bioassays. *Angewandte Chemie International Edition*, **46(8)**, pp. 1318–1320, 2007.

[4] Abe, K., Suzuki, K. & Citterio, D., Inkjet-printed microfluidic multianalyte chemical sensing paper. *Analytical Chemistry*, **80(18)**, pp. 6928–6934, 2008.

[5] Li, X., Tian, J., Nguyen, T. & Shen, W., Paper-based microfluidic devices by plasma treatment. *Analytical Chemistry*, **80(23)**, pp. 9131–9134, 2008.

[6] Li, X., Tian, J., Garnier, G. & Shen, W., Fabrication of paper-based microfluidic sensors by printing. *Colloids and Surfaces B: Biointerfaces*, **76(2)**, pp. 564–570, 2010.

[7] Dungchai, W., Chailapakul, O. & Henry, C.S., Electrochemical detection for paper-based microfluidics. *Analytical Chemistry*, **81(14)**, pp. 5821–5826, 2009.

[8] Hossain, S.M.Z., Luckham, R.E., McFadden, M.J. & Brennan, J.D., Reagentless bidirectional lateral flow bioactive paper sensors for detection of pesticides in beverage and food samples. *Analytical Chemistry*, **81(21)**, pp. 9055–9064, 2009.

[9] Li, X., Tian, J. & Shen, W., Quantitative biomarker assay with microfluidic paper-based analytical devices. *Analytical and Bioanalytical Chemistry*, **396(1)**, pp. 495–501, 2010.

[10] Li, X., Tian, J. & Shen, W., Thread as a versatile material for low-cost microfluidic diagnostics. *ACS Applied Materials & Interfaces*, **2(1)**, pp. 1–6, 2010.

[11] Wakelyn, P.J., *Cotton. Kirk-Othmer Encyclopedia of Chemical Technology*, John Wiley & Sons, Inc., pp. 19–20, 2000.

[12] Blicharz, T.M., Rissin, D.M., Bowden, M., Hayman, R.B., DiCesare, C., Bhatia, J.S., Grand-Pierre, N., Siqueira, W.L., Helmerhorst, E.J., Loscalzo, J., Oppenheim, F.G. & Walt, D.R., Use of colorimetric test strips for monitoring the effect of hemodialysis on salivary nitrite and uric acid in patients with end-stage renal disease: a proof of principle. *Clinical Chemistry*, **54(9)**, pp. 1473–1480, 2008.

[13] Daniels, G. & Reid, M.E., Blood groups: the past 50 years. *Transfusion*, **50(2)**, pp. 281–289, 2010.

[14] Sun, J., Zhang, X., Broderick, M. & Fein, H., Measurement of nitric oxide production in biological systems by using Griess reaction assay, *Sensors*, **3(8)**, pp. 276–284, 2003.

[15] Nisbet, D.R., Pattanawong, S., Ritchie, N.E., Shen, W., Finkelstein, D.I., Horne, M.K. & Forsythe, J.S., Interaction of embryonic cortical neurons on nanofibrous scaffolds for neural tissue engineering. *Journal of Neural Engineering*, **4(2)**, pp. 35, 2007.

[16] Vazquez, G., Alvarez, E. & Navaza, J.M., Surface tension of alcohol + water from 20 to 50 °C. *Journal of Chemical & Engineering Data*, **40(3)**, pp. 611–614, 1995.

Author Index

Sustainable Development and Planning V

Edited by: ***C.A. BREBBIA****, Wessex Institute of Technology, UK*

This book contains the proceedings of the latest in a series of biennial conferences on the topic of sustainable regional development that began in 2003. Organised by the Wessex Institute of Technology, the conference series provides a common forum for all scientists specialising in the range of subjects included within sustainable development and planning. It has become apparent that planners, environmentalists, architects, engineers, policy makers, and economists have to work together in order to ensure that planning and development can meet our present needs without compromising future generations.

The topics covered by the papers included in the book include: City planning; Regional planning; Social and political issues; Sustainability in the built environment; Rural developments; Cultural heritage; Transportation; Ecosystems analysis, protection and remediation; Environmental management; Environmental impact assessment; Indicators of sustainability; Sustainable solutions in developing countries; Sustainable tourism; Waste management; Flood risk management; Resources management; and Industrial developments.

WIT Transactions on Ecology and the Environment, Vol 150
ISBN: 978-1-84564-544-1 eISBN: 978-1-84564-548-5
Forthcoming 2011 / apx 900pp / apx £342.00